Continuing Evaluation of the Use of Fluorides

AAAS Selected Symposia Series

Routledge
Taylor & Francis Group

LONDON AND NEW YORK

Continuing Evaluation of the Use of Fluorides

Edited by
Erling Johansen, Donald R. Taves and Thor O. Olsen

AAAS Selected Symposium 11

First published 1979 by Westview Press, Inc.

Published 2018 by Routledge
52 Vanderbilt Avenue, New York, NY 10017
2 Park Square, Milton Park, Abingdon, Oxon OX14 4RN

Routledge is an imprint of the Taylor & Francis Group, an informa business

Copyright © 1979 by the American Association for the
Advancement of Science

All rights reserved. No part of this book may be reprinted or reproduced or utilised in any
form or by any electronic, mechanical, or other means, now known or hereafter invented,
including photocopying and recording, or in any information storage or retrieval system,
without permission in writing from the publishers.

Notice:
Product or corporate names may be trademarks or registered trademarks, and are used only
for identification and explanation without intent to infringe.

Library of Congress Catalog Card Number: 78-24745
ISBN 13: 978-0-367-02072-9 (hbk)
ISBN 13: 978-0-367-17059-2 (pbk)

About the Book

This work addresses a variety of topics that are of interest in their own right and because they bear on the evaluation of the safety of fluoridation. The latter is important because the demonstration of "no effect" from the use of fluorides is very difficult and the available data need to be periodically reexamined to reduce the theoretical amount of adverse effect that may have escaped detection.

The authors were selected because of their original contributions in the use of fluorides to prevent dental caries and in modern anesthetics, hemodialysis, and the treatment of osteoporosis. Basic studies on topics such as fluoride measurement, intake, distribution, and excretion by the kidneys and individual cells are reviewed and updated with recently obtained, previously unpublished data, and claims that fluorides cause cancer, mongolism, and idiosyncratic responses are critically reviewed. Positive response to the symposium on which this book is largely based suggests that the goal of scientific objectivity in this field has been reached in large measure.

Foreword

The *AAAS Selected Symposia Series* was begun in 1977 to
provide a means for more permanently recording and more
widely disseminating some of the valuable material which is
discussed at the AAAS Annual National Meetings. The volumes
in this *Series* are based on symposia held at the Meetings
which address topics of current and continuing significance,
both within and among the sciences, and in the areas in which
science and technology impact on public policy. The *Series*
format is designed to provide for rapid dissemination of information, so the papers are not typeset but are reproduced
directly from the camera copy submitted by the authors, without copy editing. The papers are reviewed and edited by
the symposia organizers who then become the editors of the
various volumes. Most papers published in this *Series* are
original contributions which have not been previously published, although in some cases additional papers from other
sources have been added by an editor to provide a more comprehensive view of a particular topic. Symposia may be reports of new research or reviews of established work, particularly work of an interdisciplinary nature, since the AAAS
Annual Meeting typically embraces the full range of the
sciences and their societal implications.

WILLIAM D. CAREY
Executive Officer
American Association for
the Advancement of Science

Contents

List of Figures

Chapter 4

List of Tables

Acknowledgment

Financial support for this symposium was obtained from the following sources:

American Association for the Advancement of Science-
Section R, Dentistry

The Procter and Gamble Company

University of Rochester

The assistance of Colleen Dachille and Joy Howe in the preparation of the manuscript is also gratefully acknowledged.

About the Editors and Authors

Erling Johansen is the Margaret and Cy Welcher Professor of Dental Research, professor of clinical dentistry and chairman of the Department of Dental Research at the University of Rochester. He has served as a consultant to the National Institute of Dental Research and the Bureau of Environmental Health and was formerly editor of the Journal of Dental Education *and associate editor of the* Journal for Dental Research. *A fellow of the American College of Dentists, of the International College of Dentists, and of other organizations, he has received many honors and awards; most recently, the Award of Merit from the Rochester Academy of Medicine (1977). He has taught and lectured throughout the world and is the author of numerous publications on dental caries, mineral metabolism and preventive dentistry.*

Donald R. Taves is associate professor of pharmacology and toxicology and of radiation biology and biophysics at the University of Rochester School of Medicine and Dentistry. He was a member of a subcommittee of the National Academy of Sciences Safe Drinking Water Study which examined the use of fluoride in drinking water and his principal area of specialization is the toxicology and pharmacology of fluoride.

Thor O. Olsen, assistant professor of dental research at the University of Rochester, has conducted extensive research on the chemistry of teeth and calcified structures and on mineral metabolism. He has made numerous presentations at scientific meetings on his areas of specialization, and has many years of editing experience, including serving on the Journal of Dental Education.

Warren S. Guy, clinical fellow of pediatric dentistry at Children's Hospital Medical Center in Cincinnati, has conducted research on fluoride chemistry and metabolism in addition to working in dentistry for children. His publications include several on fluorine in human blood.

Harold C. Hodge is a professor in the Department of Pharmacology at the School of Medicine, University of California, San Francisco. He has published numerous papers on toxicology and pharmacology, including "Biological Effects of Inorganic Fluorides" (with Frank A. Smith) in Fluorine Chemistry, Vol. IV *(J.H. Simons, ed.; Academic Press, 1965). He has received various awards, including the H. Treadley Dean Memorial Award from the International Association for Dental Research in 1976 and the Merit Award (1969) and the Toxicology Education Award (1975) from the Society of Toxicology.*

William J. Johnson, director of the Mayo Artificial Kidney Center and professor of medicine with the Division of Nephrology at the Mayo Clinic, has been involved in the study of calcium and phosphorus metabolism and renal osteo-dystropy, potassium metabolism, and uremic neuropathy. He is past chairman of the Minnesota State Medical Association's Committee on Dialysis and Transplantation and served on the editorial board of Nephron. *He has published over 70 papers in his field.*

Jenifer Jowsey, director of orthopaedic research at the Mayo Clinic and Mayo Foundation, has worked extensively on metabolic bone disease, particularly the prevention and treatment of osteoporosis. She received the Kappa Delta Award from the American Academy of Orthopaedic Surgeons in 1975. She is the author of some 230 publications including Metabolic Diseases of Bone *(Saunders, 1977).*

P. J. Kelly, consulting physician in orthopaedic surgery at the Mayo Clinic and Mayo Foundation, has been involved in research in the area of blood flow in bone. He is past president of the Orthopaedic Research Society and has published on the subjects of circulation and physiology of bone and bone metabolism.

Thomas M. Marthaler, professor of oral epidemiology and preventive dental medicine in the Division of Applied Prevention, Department of Cariology, Periodontology and Preventive Dental Medicine, University of Zurich Dental Institute, is also head of the University's Biostatistical Center of the Medical Faculty and supervisor of public dental health, Canton of Zurich. His areas of expertise are epidemiology and prevention of dental caries and periodontitis, biostatistics and EDP systems, and medical pathology. He is the author of some 130 publications on these and related topics.

David H. Pashley, associate professor in the Department of Oral Biology/Physiology at the Medical College of Georgia, received his Ph.D. in renal physiology. He has been studying the effects of changes in pH on metabolism and has published two articles in this area: "Functions of Renal Metabolism: Regulation of pH and Substrate Interconversion by Dog Kidney Cortex" and "Fluoride Renal Clearance: A pH Dependent Event" (Am. J. Physiol., 1973 and 1976).

Mary G. Repaske, a research assistant and graduate student in the Department of Biochemistry at the University of Wisconsin, Madison, holds a B.S. in chemistry from the University of Texas, Austin. She is studying fluoride metabolism and is coauthor of "Effects of Fluoride on Cultured Cell Metabolism" (with J.W. Suttie) in Trace Element Metabolism in Man and Animals--3 *(Proceedings of the 3rd International Symposium, M. Kirchgessner, ed., 1978).*

B. L. Riggs, chairman of the Division of Endocrinology/ Metabolism in the Department of Internal Medicine at the Mayo Clinic and Mayo Foundation, has worked primarily in the field of metabolic bone disease. He was a traveling fellow of the Royal Society of Medicine in 1973 and is a member of several professional societies.

Reidar F. Sognnaes, professor of oral biology with the School of Dentistry and of anatomy with the School of Medicine at the University of California, Los Angeles, is a specialist in oral biology and forensic science. He is current president of the International Society for Forensic Odonto-Stomatology and past president of the American Institute of Oral Biology. His publications include Advances in Experimental Caries Research *and* Calcification in Biological Systems *(AAAS: 1955 and 1960).*

John W. Suttie, professor of biochemistry at the University of Wisconsin-Madison, has studied biochemical lesions caused by fluoride ingestion, fluoride as an industrial pollutant, and fluoride homeostasis. He serves on the Fluoride Panel of the National Research Council Committee on Biologic Effects of Atmosphere Pollutants and on the NRC Nutrition Subcommittee on Fluorosis in Animals. He received the Mead Johnson Award of the American Institute of Nutrition in 1974, and is the author of over 100 publications on fluoride metabolism and on vitamin K action, anticoagulants, and the chemistry of prothrombin.

Russell A. Van Dyke, associate professor in the Department of Anesthesiology at the Mayo Clinic, is a specialist in the areas of drug metabolism and toxicity of drugs. He was previously a senior research biochemist with Dow Human Research. He is the author of some 65 publications on his fields of interest.

Gary M. Whitford, assistant professor in the Department of Oral Biology/Physiology at the Medical College of Georgia, specializes in fluoride metabolism and toxicity. His publications include "Fluoride Renal Clearance: A pH Dependent Event" and "Fluoride Absorption from the Rat Urinary Bladder: A pH Dependent Event" (Am. J. Physiol., 1976 and 1971).

Continuing Evaluation of the Use of Fluorides

Introduction

There has been a long-standing interest in both the
beneficial and harmful effects of fluoride on teeth from
which the topics of current scientific interest arise. The
fascinating detail of the historical background is covered
by R. F. Sognnaes in the first chapter of this monograph.
The next three chapters deal primarily with the beneficial
effects of fluoride. T. M. Marthhaler covers the beneficial
effects to teeth from fluoride tablets and solutions which
are swallowed and absorbed. One of the important aspects of
his review concerns the degree of topical effects from
fluoridated water after the teeth have erupted. E. Johansen
and T. Olsen review topical applications of fluoride for the
control of dental caries. They include much previously unpublished work which shows that the rotting away of teeth
with radiation therapy to the head can be avoided in cancer
patients. Their findings have been extended to patients
with high susceptibility to caries of all kinds and may
change the treatment of dental caries. Chemical management,
rather than drilling and filling, might become a reality.
J. O. Jowsey, B. L. Riggs and P. J. Kelly review the use of
fluoride in the treatment of bone disease, principally
osteoporosis. Although fluoride treatment is still the only
one to give evidence of doing more than just arresting the
progress of bone loss, it remains in the experimental stage
because of questions concerning side effects, bone strength,
and best ancillary treatment.

Use of fluoride without production of adverse effects
requires cognizance of how fluoride is distributed in the
body and of all the sources of fluoride exposure. There
have been warnings that the fluoride concentration in drinking water should be lowered to compensate for an increasing
intake of fluoride from food. The key in determining the
validity of this claim, as well as understanding how fluoride
behaves in the body, depends on the accurate analysis of

fluoride in food and in body fluids or tissues. W. S. Guy's
chapter has a historical review of the analytical determination of fluoride in one tissue, blood. A dramatic decrease
in the accepted value over the past century and good agreement by several methods in the past ten years make us pain-
fully aware of how careful scientists need to be in drawing
conclusions from data which has not been confirmed by independent tests. The question of an increasing intake of
fluoride from food is covered by D. R. Taves. Data are
presented which show that faulty analytic methods were
involved and three types of evidence are presented which
indicate that no increase in fluoride intake has occurred.
How fluoride behaves in the body is discussed by Taves and
Guy in the chapter on the distribution of fluoride in the
various body compartments. Contrary to some investigators'
opinions, they conclude that serum fluoride is directly
related to the concentration in other compartments and should
prove valuable in evaluating risk of toxicity. Changes in
the efficiency of the kidneys in excreting fluoride might
be important in providing optimum serum fluoride levels in
the face of markedly different intakes of fluoridated water.
G. M. Whitford and D. H. Pashley consider various factors
which might change the efficiency of renal excretion, and
they conclude that the important variable is pH, which is
increased by dilution (i.e., by increased fluid intake).
They present new evidence which indicates that pH is important in the susceptibility to toxic effects of fluoride over
and beyond its effect on the excretion of fluoride. The two
effects together promise to make treatment of acute fluoride
poisonings much more effective than it has been. Fascinating work on fluoride at the cellular level is reviewed and
extended by M. G. Repaske and J. W. Suttie. The indication
that cells can differ markedly in the ability to pump
fluoride from the inside to the outside may provide verification and explanation of idiosyncratic responses to fluoride. R. A. VanDyke reviews how fluorinated anesthetics are
metabolized by subcellular components. This has become
important not only in the use of fluorinated drugs but
because organic fluoride, presumably from industrial contamination, has been found to be generally present in human
serum. Also, because of the release of inorganic fluoride,
the studies on anesthetics have resulted in a much better
understanding of the effect of fluoride on the kidney.

Three chapters deal primarily with adverse effects.
H. C. Hodge reviews the development of dose schedules for
the prevention of caries with supplemental fluoride which,
under certain circumstances, can lead to undesirable levels
of dental mottling. W. J. Johnson, D. R. Taves and J. O.
Jowsey conclude from the literature and from patient

histories at Mayo Clinic that patients with renal disease
are sometimes affected by fluoridated hemodialysate or from
drinking water with a natural content of more than 2 ppm
fluoride. Therefore, they recommend that patients with
long-term renal disease be monitored for excessive build-up
of fluoride in the serum. D. R. Taves covers the wide
variety of claims of harm from fluoridation and concludes
that these have not been demonstrated adequately from a
scientific point of view, but notes that more work is needed
before assuming that these claims are baseless.

Benefit relative to risk has not been considered for
two reasons. First, the units used in measuring the effects
are different, therefore, judgement about the importance of
dissimilar interests cannot be tested scientifically.
Second, the scientist has a special role to play in continuing to evaluate and refine the data base on which that subjective decision rests, and this function can best be served
by scientists who are open-minded regarding new evidence
which lends weight to either side. The need for continuing
evaluation of fluoridation stems from the fact that scientifically we cannot prove safety, we can only narrow the
magnitude of effects that might be missed.

The Editors

Historical Perspectives

Reidar F. Sognnaes

This state, Colorado, and this organization, the American Association for the Advancement of Science, have done much to pioneer and publish the historical developments of fluoridation. Therefore, it is quite appropriate that we meet here to observe the 60th anniversary of the discovery of the so-called "Colorado brown stain" -- later to become known as mottled teeth or dental fluorosis -- and to update the publications on fluoridation through the forum of AAAS; and in retrospect, we must also pay tribute to some of the early observers who had stimulated interest in the subject many years before.

Some Pre-Colorado Observations

Even before the turn of the century, according to a dental research conference held in Germany, one observer (Kühns, 1888) had reported a darkish tooth mottling in an entire family which had lived in Durango, Mexico. Better known, in retrospect, is a publication at the turn of the century by Eager (1901; 1902), who attributed the unusual darkened color of the teeth of Italian immigrants to the drinking water contaminated from some volcanic source in their place of birth and youth. Though fluoride in the drinking water was not scientifically related to mottled enamel until 1931, it was suggested more than 100 years earlier by Berzelius (1822) that the fluoride which he had been able to demonstrate in teeth was probably brought there through drinking water, in which he had demonstrated the presence of over 1 part per million of fluoride. Before Berzelius, the demonstration of fluoride in dental hard tissues was actually made by Morichini (1805), first in an elephant tusk, i.e., in ivory, and then in human dental enamel.

Even in terms of control of dental caries, the potential significance of fluoride was conceived as a possibility during the past century. For example, Crichton-Browne (1892) went so far as to suggest that dental caries was a sign of nutritional deficiency of fluoride. Also, on the basis of anti-caries experiments of 100 years ago, Erhardt (1874) suggested that children and women should eat fluoride-containing pills. A number of recent investigators have done sophisticated work on the fluoride content of teeth, normal and decayed. But before the turn of the century, two studies (Wrampelmeyer, 1893; Hempel and Scheffler, 1899) reported results suggesting a higher fluoride content in intact than in decay-susceptible teeth. Other authors of the same period speculated that the fluoride content of enamel had a caries-inhibitory effect. One writer, Michel (1897), even became so specific as to suggest that caries was related to a fluoride effect on the adjacent microflora of the tooth surface. And then, to complete this ante-fluoridation period, there was a European report by a fluoridation enthusiast of seventy years ago, Deninger (1907), who went beyond speculation. He adopted in a practical manner the idea of giving some of his patients tablets containing calcium fluoride in order to control their susceptibility to dental decay.

Earliest Colorado Observations

To obtain a proper perspective, it is necessary to make comparisons between the earliest and latest discoveries. Dental lesions comparable to the 1916 observations on "Colorado brown stain" have actually been alluded to under different names from various parts of the world prior to that time. Six years earlier, right here in Colorado, a report by Fynn (1910) indicated that a majority of children born and raised in Colorado Springs had mottled teeth; but he ruled out drinking water as a potential cause. Shortly thereafter, Rodriguez (1915), who worked as a dentist with the United States Indian Service, reported on brown stains of the enamel among the Pima Indians of Arizona, and he did conclude that the condition might be due to the nature of the water supply. According to that author, Dr. Fred McKay of Colorado evidently had visited the Pima reservation shortly before presenting the famous 1916 report on the subject in collaboration with Dr. G.V. Black. Their studies will be referred to under the next heading.

Early Research on Fluoride and Dental Caries in Man and Animals

Human Studies. Black (1916) and McKay (1916) published the first detailed studies of mottled teeth, an endemic developmental imperfection of the enamel now more properly called dental fluorosis. Black's histological studies were continued by Williams (1923) and Beust (1925), who found that not only the enamel's interprismatic substance was affected, but also the enamel rods, as well as the dentin to some extent.

Black (1916) and McKay (1916, 1928, 1929), Bunting et al. (1928) early called attention to the fact that mottled teeth, in spite of their imperfect structure, were relatively resistant to dental caries. What in retrospect may have been one of Dr. McKay's most important, yet less well known, reports was written after he had moved his dental practice from Colorado to New York (McKay, 1926), when he openly implicated local water supplies as the probable cause for endemic mottling of teeth. There followed a flood of letters to the editor expressing a mixture of congratulations, concern and critiques, and commenting on a variety of chemical culprits potentially involved. Only one letter, written by a water plant chemist from Toronto, Ontario (Hannon, 1926), happened to allude to the possible role of water fluoride: "Of the mineral elements at present known to be common to both water and enamel, the chief are calcium, phosphorus, and fluorine... But when we consider fluorine, all is at present shrouded in obscurity... I must confess complete ignorance." But alas, the writer evidently did not think in terms of a fluoride overdose, but rather a deficiency: "... should the incriminated waters prove to be all alike fluorine-free, the case for fluorine deficiency will become strong; ...". In 1931, three independent studies by Smith, Lantz and Smith (June, 1931), Churchill (September, 1931) and Velu (November, 1931) pointed to fluorine as the cause of mottled enamel, findings that were subsequently confirmed by McKay (1933), Dean, McKay and Elvove (1938) with their successful prevention of mottled enamel by changing, in endemic areas, to a water supply containing less fluorine. Masaki (1931), Ainsworth (1933), Erasquin (1935), Arnim, Aberle and Pitney (1937), Dean (1938) and Dean et al (1939) added support for the contention that fluorosed teeth (or children in endemic areas) present a certain resistance to the attack of dental caries.

The last authors (Dean, 1938; Dean et al., 1939) particularly related this to the use of water with a relatively high fluorine concentration, and found that the caries resistance also holds true for those teeth which do not show clinical evidence of fluorosis.

Animal Studies. In the early thirties Hoppert, Webber and Caniff (1931) - rather accidentally - produced a high frequency of experimental caries in white rats fed a coarse, presumably adequate, corn diet. This gave rise to a number of caries studies in rats, in the beginning especially related to the cause and nature of the lesions (Klein and McCollum, 1931; Rosebury, Karshan and Foley, 1933; Bibby and Sedwick, 1933; and King, 1935). Later Lilly (1938) obtained a considerable reduction in the rat caries when casein replaced the powdered milk of Hoppert, Webber and Canniff's caries-producing diet.

Hodge, Luce-Clausen and Brown (1939) by spectroscopic analysis demonstrated a fluorine content of 0.1% in commercial casein [i.e., 1,000 ppm], enough to produce fluorosis in rats. Subsequently Hodge and Finn (1939) showed that reduction of experimental rat caries may be obtained by fluorine-contaminated casein, and not by chemicaly pure casein, when added to a caries-producing diet.

In the meantime, Miller (1938), although his material was smaller, had for the first time demonstrated an inhibition of experimental rat caries by fluorides (as well as iodoacetic acid) incorporated in a similar, coarse, caries-producing diet. Fluoride deficiency is another story, reopened recently in another context.

Current Status of Water Fluoridation

As we return to Colorado today the evidence is clear. It was the water. The elevated fluoride content of the Colorado water was confirmed by Boissevain (1933). By proper adjustment of fluoride levels (Cox, 1939), both dental mottling and dental decay may be prevented. Now a report from the Colorado Department of Health (Bruck, 1971) has commented on the great progress that has been made during the thirty years since endemic research began on the "Colorado brown stain". That overdose problem has been solved and now the majority of the population has an appropriately adjusted fluoride intake. This is the result of cooperation between citizens generally and the dental, medical, public health and government officials.

The American Association for the Advancement of Science showed early leadership in publishing several symposium monographs on advances in fluoridation research between the early forties and the mid-fifties (Moulton, 1942, 1946 and Shaw, 1954). Research on fluoride chemistry and toxicology has been thoroughly evaluated (Hodge and Smith, 1965, 1968) and the important contributions by the U.S. Public Health Service, during a whole generation of national fluoridation efforts, have been proudly acclaimed in Fluoridation - the Search and the Victory (McClure, 1970). There is an outstanding monograph in Danish dealing with dental fluorosis and caries research (Möller, 1965).

For public edification, one account, Fifty-two Pearls and their Environment, has enthusiastically dealt with fluorides and other newer means of caries prevention (Muhler, 1965). To update the education of dentists and dentists-to-be, there is now a concise textbook on Fluorides and Dental Caries (Newbrun, 1975, 1977). And, most recently, another large monograph is devoted primarily to topical fluoridation, based on a USPHS supported workshop, entitled Cariostatic Mechanisms of Fluorides (Brown and König, 1977).

Beyond the more comprehensive symposium monographs and texts on fluoridation, there have also appeared a number of substantial single-authored reviews. From these I might mention, in celebration of our American fluoridation milestones, one from the Eastern and one from the Western U.S.A. Thus, at the time of the 20th anniversary of the first water fluoridation trials (initiated in 1945 in Grand Rapids, Michigan and Newburgh, New York), an excellent account on the status of water fluoridation was published in the Medical Progress series of the New England Journal of Medicine (Dunning, 1965). And then, five years later, celebrating the 25th anniversary of the same event, Knutson (1970) gave an account of his 1945 journey with Dr. Trendley Dean, then NIH's Mr. Water Fluoridation himself, from Washington, D.C. to Grand Rapids, Michigan, in order to help initiate one of the first two field trials of water fluoridation in the United States. Simultaneously, with equally justifiable pride, Canada took note of its own 25th anniversary of water fluoridation in Brantford, Ontario (Connor, 1970). Three years later a paper was published on a dozen Mexican localities having natural fluoride levels between 1 and 3 ppm F (Sanchez Y Castillo and Gomez-Castellanos, 1973).

The latest figures from the Center for Disease Control in Atlanta show that out of the more than 213 million people served by public water supplies in the U.S. nearly 106 million are now receiving fluoridated water. The percentage of the population in each state receiving fluoridated water in the early 1970's is given in Table 1. The first eight states to pass mandatory fluoridation legislation were Connecticut, Michigan, Georgia, Minnesota, South Dakota, Nebraska, Illinois and Ohio. Statewide fluoridation referenda were passed in Washington and Oregon in November 1976, but defeated in Utah. Among the larger American cities, Los Angeles, Portland, New Orleans have hitherto failed to adopt water fluoridation; and in Boston fluoridation is not as yet activated though it is approved and imminent at this writing. In all of the United States, about one half of the total population (46%) is receiving fluoridated water (Table 1).

Worldwide Water Fluoridation

Outside of the United States and Canada, the Western nations have been slow in adopting water fluoridation as a public health measure. One of the earliest countries to move ahead was New Zealand (Ludwig, 1971) which has reported significant caries control since 1954. A few years later, in Australia (Barnard, 1969), New South Wales passed legislation permitting boards of health to take action in the matter.

From Western Europe there have been reports on remarkable caries rate reductions after fifteen years of water fluoridation in the Netherlands (Flaumenhaft, 1971). In Eastern Europe, between 1953 and 1967, water fluoridation was implemented in pilot cities in Czechoslovakia (Kostland and Jiraskova, 1967), East Germany (Kunzel, 1970, 1970a, 1974) Romania (Benedek, 1964; Csögör, Gozner and Cristoloveanu, 1968), Poland (Wigdorowiez-Macowerowa, 1969) and USSR (Rybacov 1968; Gabowitsch and Owrutzki, 1969; Basiyan, 1970), as well as in Bulgaria (see Kunzel, 1967, 1970a).

In Eastern Europe, once water fluoridation showed encouraging results, implementation was apparently taken for granted. For example, in Poland ten years ago a so-called scientific-technical fluoride prophylaxis committee was formed to find means for fluoridation of public waters. Within half a year the city of Wroclow with half a million people was fluoridated. Two years later, the Ministry of Health and Welfare recommended nationwide fluoridation of drinking waters (Wigdorowica-Makowerowa, 1969; Wigdorowicz-Makowerowa, Potoczek and Rakowicz, 1970).

To summarize, by the time of the 25th anniversary of water fluoridation in the United States and Canada, there were over thirty countries around the world which had begun to implement this public health measure (Bernhardt, 1970). Among these were eight nations of Central and South America, i.e., Brazil, Chile, Colombia, El Salvador, Panama, Paraguay, Peru and Venezuela. In Western Europe another eight countries had begun fluoridation: Belgium, Great Britain, Ireland, Finland, the Netherlands, Sweden (since discontinued), Switzerland and West Germany. In Eastern Europe, Czechoslovakia, East Germany, Poland, Romania and USSR are very actively pursuing water fluoridation. And in the Far East, fluoridation programs have been implemented in Japan, China, New Guinea, Hong Kong and Singapore (Bernhardt, 1970). There is limited information on fluoridation in the Near East and Africa, where some countries have a very low dentist-population ratio.

The World Health Organization published a comprehensive multi-authored report on Fluorides and Human Health in 1970, and now, after we have passed the 30th anniversary of water fluoridation in the United States, has launched a new world-wide fluoridation effort (WHO, 1975). Based on the latest research and practical experiences on the benefits and safety of fluoridation, a new directive by the WHO Director General is restating its promotion of fluoridation and its preparedness to coordinate and stimulate the efforts of member nations who wish to establish national programs for fluoridation.

Bottlenecks to Public Water Fluoridation

Many of those who were involved in the implementation of water fluoridation a generation ago optimistically predicted that the prospect for nationwide fluoridation within a five-year period was to be expected. But evidently such an outlook was based on unreasonable expectations regarding both professional and public understanding of the scientific, social and political issues involved (Dunning, 1965).

Here in the United States, the public referendum has been the most common approach to introducing water fluoridation. This has attracted a great deal of interest and debate inside and outside the scientific and professional community. In fact, the use of referenda has greatly delayed water fluoridation in many localities (Frankel and Alluklan, 1973). Massachusetts had sixteen fluoridation referenda before a new law was enacted

TABLE 1   PERCENT OF POPULATION OF UNITED STATES AND PUERTO RICO SERVED WITH NATURAL AND CONTROLLED FLUORIDATED WATER

State	Percent of total population served by fluoridated water	Percent of population with public water supplies served by fluoridation
Alabama	25.9	41.0
Alaska	44.2	80.5
Arizona	17.5	20.0
Arkansas	37.0	68.5
California	18.4	19.6
Colorado	73.8	90.0
Connecticut	73.2	90.9
Delaware	40.5	54.6
District of Columbia	100.0	100.0
Florida	29.0	37.5
Georgia	47.7	68.5
Hawaii	12.8	13.3
Idaho	17.9	27.5
Illinois	85.3	99.0
Indiana	59.4	89.3
Iowa	56.2	81.7
Kansas	45.3	59.3
Kentucky	46.4	81.2
Louisiana	8.0	10.6
Maine	38.9	62.0
Maryland	77.5	98.6
Massachusetts	14.9	15.6
Michigan	64.1	91.2
Minnesota	74.3	99.0
Mississippi	22.5	44.6
Missouri	45.6	61.9
Montana	19.9	29.5

State	Percent of total population served by fluoridated water	Percent of population with public water supplies served by fluoridation
Nebraska	48.2	67.4
Nevada	3.3	3.9
New Hampshire	11.6	17.6
New Jersey	13.1	14.8
New Mexico	40.2	55.5
New York	67.0	76.3
North Carolina	38.5	75.6
North Dakota	48.6	92.4
Ohio	42.3	54.4
Oklahoma	57.0	75.5
Oregon	16.5	21.7
Pennsylvania	43.5	53.4
Rhode Island	81.4	90.9
South Carolina	36.6	65.0
South Dakota	53.5	92.0
Tennessee	44.3	67.9
Texas	51.4	63.8
Utah	2.5	2.7
Vermont	26.6	45.7
Virginia	62.3	97.0
Washington	38.8	46.6
West Virginia	51.5	84.2
Wisconsin	62.0	95.5
Wyoming	30.0	40.6
Puerto Rico	67.5	93.7
ALL OF UNITED STATES	46.0	59.2

in 1968, which authorized the local boards of health to make decisions in the matter.

Ohio is credited with an exemplary approach to implementation of communal water fluoridation. It began in February 1969 with the aid of a citizen committee in addition to scientific and professional support. After labor management and state health and welfare departments spoke up in favor of fluoridation, the governor signed a bill to fluoridate all of the state's water supplies serving more than 5,000,000 people (ADA Council Report, 1970).

Other parts of the Western World have not been spared the water fluoridation polemics. In the United Kingdom and the Scandinavian countries -- suffering from some of the highest dental decay rates in the world -- water fluoridation programs have been bogged down by widely publicized polemics. The British have been particularly concerned with the appropriateness of the process involved in local government decisions regarding fluoridation (Dickson, 1969; Brien, 1970; Jackson, 1972).

From West Germany a pessimistic report (Naumann, 1970) concluded that the German federal government would be unlikely to issue a general authorization for fluoridation. In the Netherlands the Minister of Public Health must be petitioned by a community wishing fluoridation; and he has recommended that for those who demand it the communities must provide unfluoridated water. Austria has had emotional propaganda even against individual fluoridation, the so-called "fluoride tablets war" (Hurny, 1974).

Based on a review of 23 articles on actual case histories, Petterson (1969, 1971, 1972) has emphasized the crucial influence of elected officials on decisions to fluoridate. For example, after the Swedish parliament repealed the enabling legislation on water fluoridation in 1971, there has been expressed concern, for example in decay-prone Britain, that a similar situation could arise in other countries unless the political side of the fluoridation issue is treated more seriously (Burt and Petterson, 1972).

It is interesting to note that in absence of water fluoridation our Swedish colleagues have done some of the most extensive and productive research on the use of other fluoride procedures. For example, large groups of Swedish school children have participated in supervised toothbrushing and mouth washing with fluoride solutions (Berggren and Welander, 1960; Torell and Siberg, 1962).

Among Swedish professionals favoring fluoridation, a number of recommendations have also been made on how to improve the public's understanding of and attitude towards this issue, i.e., how to minimize the polemics of the opposition. Unfortunately, inadequate information on fluoride does not only exist among the public, which may be understandable, but according to some Scandinavian studies, even among dental students (Petterson, 1971).

In a cluster sample, designed to be representative of the some three million Norwegians over fifteen years of age, only 34% were in favor of fluoridation, whereas the majority either opposed it (36%) or did not have an opinion (30%). It was found that more of the younger population agreed to fluoridation, especially those in higher income brackets living in the cities. But for the moment it is concluded that there is not a sufficient attitudinal basis for the introduction of water fluoridation in Norway (Helöe and Birkeland, 1974).

Removal of Water Fluoride

Once the mottling effect of an overdose of naturally occurring fluorides was confirmed, changes in the source of drinking water solved that problem in most places. Therefore, today there is no great research interest in defluoridation. Yet there are still localities, for example in India, where people suffer from excessive natural fluoridation of their water supplies, and consequently associated problems of fluorosis.

An entirely different story relates to the discontinuation of fluoridation of communal water supplies. There are now a few noteworthy reports on such cases, showing a significant detrimental effect on children's teeth. In Kilmarnock, Scotland (anon., 1969) and in Antigo, Wisconsin (Lemke, Doherty and Arra, 1970), it was discovered that the dental deterioration became evident in the non-fluoridated crop of teeth erupted within five years after the water fluoridation was discontinued. With such a dramatic reconfirmation of benefits, the Wisconsin community decided to reinstate fluoridation in 1965.

School Drinking Water Fluoridation

In rural school districts where water fluoridation cannot be implemented through communal water supplies, school water fluoridation has been demonstrated to be successful in controlling dental caries of children (Horowitz, Heifetz and Law, 1972 and Heifetz and Horowitz, 1974). Because the children, while in school, only ingest a portion of their total water intake, the school water was adjusted to from four to six times the usual level of fluoride, namely to from 4 to 6 ppm F.

Sources of Fluoride Other than Drinking Water

Natural Food Fluorides

The major dietary sources of fluoride are tea and fish. In older children and adults tea can become one of the more significant sources of fluoride. The pros and cons of this fluoride source have understandably received special attention in England (Cook, 1969; Jenkins, 1969). Japanese investigators, with their flair for quantitative data, have suggested that a high concentration of tea fluoride actually may serve as a criterion for selecting good tea leaves (Kondo, Yasaki and Okudera, 1973). It has been calculated that Japanese adults obtain from 0.48 to 0.97 mg per day of fluoride from their green tea.

In Norway, a recent study measured the fluoride extractable from a variety of commercially available tea bags (Ulvestad, 1973). It was found that after three minutes of brewing with non-fluoridated water, the average fluoride concentration was from 0.90 to as high as 2.25 ppm. In other words, a daily uptake of 1 mg F could not be achieved without drinking at least 4 to 5 cups of tea per day, hardly a custom of Norwegian children.

On the other hand, a favorable topical effect of tea on teeth might be of some significance, both in youth and adults. This would be particularly true if the tea is touched up with some sour lemon, which would render the tooth surface more reactive to fluoride. At least in one experiment, on rats, Anaise et al. (1974) found that drinking of tea with lemon juice caused a demonstrable increase in fluoride uptake in the enamel surface. We do not have detailed data on caries prevention in human tea drinkers.

The potential significance of the elevated fluoride content in fish became of special interest to this author some forty years ago, when he observed the extremely high fish consumption on the island of Tristan de Cunha. (Sognnaes, 1941a), coupled with the low caries frequency and relatively high fluoride content of their teeth

(Sognnaes and Armstrong, 1941). Since then, there has been increasing interest in fluoride in foods generally, including fish (Birch, 1969; Petersen, 1970; San Filippo and Battistone, 1971; San Filippo, et al., 1972), oriental fish sauce (Kridakorn, 1973) and fish protein concentrate (anon., 1970; Stillings et al., 1973 Kerley, 1973). This protein-fluoride combination, may well merit renewed attention, both on theoretical and practical grounds, for it is often lost sight of, both by our profession and patients, that the organic matrix of a healthy protein-aceous material may be vitally important in the deposition of fluoride and the other hard tissue elements of an optimal tooth and bone structure.

Fluoride Supplementation of Food

Dietary elements considered for fluoride supple-mentation have included milk, bread, cereal and salt (Sognnaes, et al., 1953; Sognnaes, 1954; Birch, 1969; Möller, 1969). Most trials in this category have been with milk (Schmidt, 1972; Stamm, 1972; anon. 1975) and with salt, namely in Switzerland, Hungary and Colombia (Braun, 1971; Toth, 1973; Mejia, Espinal and Velez, 1972).

Milk. There is presently a good deal of interest in milk fluoridation because both human and bovine milk are very low in fluoride. It is now quite clear that some bottle-fed babies and their developing teeth are receiving supplementary fluoride through the formulas mixed with fluoridated water (Ericsson and Ribelius, 1970, 1971; Ericsson, 1973). Because milk is low in fluoride, yet represents the bulk of the fluid intake for many children, it is only natural that milk fluoridation has been explored. Even in a fluoridated community, fluoridation of milk to contribute from 1/2 to 1 mg of fluoride per day, has been suggested for youngsters between one and six years of age (Martin, 1972).

Salt. A Bulgarian study by Docev and Kalajdziev (1973) suggests that the natural mineral water from its Ovosniste village, containing 24 ppm fluoride, should be evaporated to a salt concentrate and utilized for mass administration, at three levels of increasingly concen-trated dosages, ending with the dried residue. Another paper from Bulgaria, by Atanasov, Karaivanova and Papazjan (1975), suggest that in the district of Burgas there may be a potentiation of the caries-preventive effect of fluoride, due to a synergistic action with other trace elements, such as vanadium, molybdenum and magnesium. From Italy has appeared an enthusiastic report by Lukacs (1972) that evaporation of sea water could provide an ideal dose of fluoride with the average daily pinch of sea salt being used in the home.

Fluoride Tablets and Drops

It is not surprising that fluoride supplemented medications, including vitamin preparations, have been explored both here (e.g., Hennon, Stookey and Muhler, 1969, 1972) and abroad (e.g., Bervenmark and Hamberg, 1974). The potential of these preparations continues to be under active investigation and current results will be discussed in other sections of this monograph.

Is 1 ppm of Fluoride Enough?

In the past one might well have asked "Why should drinking water be used as a fluoride vehicle if this preventive measure is only of significance to growing children?" To this question one may now rightfully reply that even higher levels of fluoride might have preventive effects with regard to skeletal diseases in the adult. This suggestion is based on observations that in communities with elevated fluoride levels, of 4 ppm or so, there has been roentgenological evidence of less osteo-porosis, and apparently also less atherosclerosis as well (Bernstein et al., 1966). In other words, it may well be that the amount of fluoride provided by water fluoridated at the 1 ppm level may be too little rather than too much for long term preventive effects on the skeletal and vascular systems in adult man (Sognnaes, 1967).

In the treatment of bone diseases physicians have prescribed daily intakes of fluorides more than twenty times the level recommended for preventive dentistry (Sognnaes, 1965). Medical aspects are treated later.

Topical Fluoride Applications

A large proportion of recent research has dealt with the topical effect of fluoride on dental caries. Several major monographs have been largely devoted to this subject (Forrester and Schulz, 1974; Brown and König, 1977). Also, it will be dealt with in detail in discussions to follow this presentation. The development of the concept and the first in vivo test of the topical application approach to caries control was almost exclusively a product of the small but enthusiastic pre-World War II dental research group at the School of Medicine and Dentistry of the University of Rochester. The initial laboratory research which created the basis for topical applications was documented by one of those investigators (Volker, 1942) in the first AAAS symposium monograph on caries and fluoride (Moulton, 1942).

Subsequent wartime research was summarized by Knutson and Armstrong (1944) and in a postwar AAAS monograph (Moulton, 1946), both with respect to the basic experimental observations (Hodge and Sognnaes, 1946) and the initial clinical field trials (Bibby, 1946).

Before considering the current status of topical fluorides it should be pointed out that this aspect of the fluoride topic, perhaps more than any other, requires careful, continuous evaluation for reasons that should become apparent, though not generally recognized.

It was natural to reason, during the early experimental work on topical fluoride application, that the highest possible concentration would be necessary to modify the chemical composition of the seemingly solid dental enamel. In fact, to achieve a high level the present author chose one of the most soluble salts, namely potassim fluoride, for the first topical tests in animals (Sognnaes, 1941). For the clinical trials, on the other hand, this author is not sure that adequate consideration was given to the wisdom of using solutions containing some 10,000 ppm of fluoride (i.e., 2% NaF) and more. Not only would this high fluoride concentration come in contact with the teeth, as intended, but also with the gingival and subgingival supporting tissues, as well as adjacent mucous membranes of the lips, cheeks, palate, tongue and "thin-skinned" floor of the mouth.

We need to explore both the local topical soft tissue influences and the systemic effects, intended or not, as indicated by several investigators, for example, Douglas (1957), Lange et al. (1971) and Gaum, Cataldo and Shiere (1973). There are also somewhat enigmatic reports on the direct effect on (a) adjacent periodontal soft tissues (e.g., Lindhe, Hansson and Branemark, 1971; Lourides et al., 1971; Mieler, Kittler and Mieler, 1972; Poulsen and Meller, 1974); (b) the dentinal pathways to the pulp tissue, with and without a varnish vehicle (Heuser, Schmidt and Krumme, 1970; Tsutsui, 1970; Myers et al., 1971; Furseth and Mjör, 1973); and (c) on periodontal wound healing (Hars and Massler, 1973; Librus et al., 1973) -- i.e., conditions which may expose tens of thousands of intraoral capillaries to tens of thousands of parts per million of fluoride.

There should be continuing concern and control with fluorides in all forms that are now becoming individually administered for home care (tablets, mouthwash, gels, toothpaste, etc.). Even though there has been no vocal opposition, but rather world-wide applause for these approaches from water fluoridation opponents, it is the duty of the profession to explore further the scientific and clinical merit of the increasing variety of dosages and vehicles now being proposed. The high concentrations of some products may be neither biologically desirable nor clinically necessary.

Fluoridated Fillings

There have been a number of innovative efforts to couple high fluoride concentrations to various more or less conventional dental materials, such as varnish (e.g., Koch and Petersson, 1975; Maiwald, 1974), dental fissure sealants (e.g., Gwinnet, 1973), adhesive dental cements (e.g. Myers et al., 1973), as well as to the old restorative standby, amalgam (e.g., Kota, Katoh and Hosoda, 1973). These fluoride preparations have been created to prevent decay in the hard tissues adjacent to dental restorations and appliances. Unfortunately, little if anything is actually known regarding reactions of the soft tissues in the mucous membranes of the lips and cheeks to prolonged direct contact with these fluoride-fortified dental products.

Concluding Comments

In retrospect, there is no other development in modern dentistry comparable to fluoridation, which emerged as a result of pioneering studies in Colorado sixty years ago and has now spread to some thirty countries on four continents. Nearly all other major milestones in the history of dentistry have been primarily related to technological and clinical advances in restoring the generally defective dentition of modern man. In addition to its significance as a public health measure, fluoridation also has been salutary to dental research in general and, especially in the United States, has resulted in greater federal support, which was virtually non-existent prior to the fluoridation breakthrough. The impact on education and in turn on dental and medical practice is also slowly but surely becoming apparent, as innovative oral fluoride applications beyond the conventional public water fluoridation and dental office topical applications are being developed.

References

American Dental Association Council on Dental Health and Council on Legislation (1970) Ohio - a model fluoridation campaign for state or community. J. Amer. Dent. Assoc. 80:814-817.

Ainsworth, T. (1933) Mottled teeth. Brit. Dent. J. 55:233-250; 274-276.

Anaise, J., Gedalia, I., Draingel, A. and Westreich, Y. (1974) Fluoride uptake in rats given nonacidulated and acidulated tea. J. Dent. Res. 53:140.

Anonymous (1969) Fluoridation - the debate continues Roy. Soc. HeaLth J. 89:260.

Anonymous (1970) The significance of fluorides in fish protein concentrates. Nutr. Rev. 28:235-236.

Anonymous (1975) Alternatives to the fluoridation of water. British Med. J. 1: 535-536.

Arnim, S.S., Aberle, S.D., and Pitney, H. (1937) A study of dental changes in a group of Pueblo Indian children. J. Amer. Dent. Assoc. 24:478-480.

Barnard, P.D. (1969) Communities fluoridated in Australia, 1968. Austral. Dent. J. 14:392-395.

Basiyan, G.V. (1970) Dental caries in transpolar conditions (Murmansk school children). Stomatologia (Moskwa) 49:4, 78.

Benedek, J. (1964) Date asuprainfluentei cariopreventive a fluorizarii apei potabile din Tirgu Mures. Stomatologia (Bukarest) 11:213

Bernhardt, M.D. (1970) Fluoridation international. J. Amer. Dent. Assoc. 80:731-734.

Bernstein, D.S., Sadowsky, N., Hegsted, D.M., Guri, C.D. and Stare, F.J. (1966) Prevalence of Osteoporosis in high-and low-fluoride areas in North Dakota. J. Am. Med. Assoc. 198(No. 5): 499-504.

Berggren, H., and Welander, E. (1960) Supervised toothbrushing with a sodium fluoride solution in 5000 Swedish school children. Acta Odont. Scand. 18:209-254.

Bervenmark, H., and Hamberg, L. (1974) Fluorine concentrations in deciduous human teeth after oral administration of sodium fluoride in vitamin solution. Acta Paediat. Scand. 63:232-234.

Berzelius, J. (1822) Extrait d'une lettre de M. Berzelius a M. Berthollet. Ann. Chim. Phys. 21:246-249.

Beust, T.B. (1925) A contribution to the etiology of mottled enamel. J. Amer. Dent. Assoc. 12:1059-1066.

Bibby, B.G. (1946) Topical applications of fluorides as a method of combating dental caries. In Dental Caries and Fluorine (F.A. Moulton, editor) Amer. Assoc. Adv. Science, Washington, D.C., p. 93-98.

Bibby, B.G., and Sedwick, H.J.   (1933)   Formation of cavities in the molar teeth of rats.   J. Dent. Res. 13:429-441.

Birch, R.C.   (1969)   The role of dietary supplements of fluoride in dental health programs for fluoride-deficient areas.   J. Public Health Dent.   29:170-187.

Black, G.V.   (in collaboration with McKay, F.S.)   (1916) Mottled teeth: an endemic developmental imperfection of the enamel of the teeth heretofore unknown in the literature of dentistry.   Dental Cosmos 58:129-156.

Boissevain, C.H.   (1933)   The presence of fluoride in the water supply of Colorado and its relation to the occurrence of mottled enamel.   Colo. Med. 30:142-148.

Braun, F.   (1971)   Fluoridation of domestic salt.   Öffl. Gesundheitswesen 33:182-183.

Brien, A.P.   (1970)   The decision process in local government: a case study of fluoridation in Hull. Pub. Admin. 48:153-168.

Brown, W.E. and König, K.G. (editors) (1977) Cariostatic Mechanisms of Fluorides.   Proceedings of a Workshop Organized by the American Dental Association Health Foundation and the National Institute of Dental Research, Naples, Fla., 1976.   Caries Research. 11 (suppl. 1):1-327.

Bruck, T.   (1971)   Water fluoridation as a preventive health measure.   J. Colo. Dent. Assoc.   49:14-16.

Bunting, R.W., Crowley, M., Hard, D.G., and Keller, M. (1928)   Further studies of the relation of Bacillus Acidophilus to dental caries.   Dental Cosmos. 70:1002-1009.

Burt, B.A., and Petterson, E.O.   (1972)   Fluoridation: Developments in Sweden.   Brit. Dent. J.   133:57-59.

Churchill, H.V.   (1931)   Occurrence of fluorides in some waters of the United States.   Ind. Eng. Chem. 23:996-998.

Connor, R.A.   (1970)   Twenty-fifth anniversary of fluoridation.   A public health success story.   Canad. J. Public Health.   61:283-284.

Cook, H.A. (1969) Fluoride and tea. Lancet 2:329.

Cox, G.J.   (1939)   New knowledge of fluorine in relation to dental caries.   J. Amer. Water Works Assoc. 31:1926-1930.

Crichton-Browne, J. (1892) An address on tooth culture. Lancet 17: II:6-10.

Csögör, L., Gozner, N., and Cristoloveanu, L.   (1968) Profilaxia cariei dentare prin fluorizarea apei potabile in Tirgu Mures.   Stomatologia (Bukarest) 15:33.

Dean, H.T. (1938) Endemic dental fluorosis and its relation to dental caries. U.S. Public Health Reports. 53:1443-1452.

Dean, H.T., Jay, P., Arnold, F.A., McClure, F.J., and Elvove, E. (1939) Domestic water and dental caries, including certain epidemiological aspects of oral L. Acidophilus. U.S. Public Health Reports. 54:862-888.

Dean, H.T., McKay, F.S., and Elvove, E. (1938) Mottled enamel survey of Bauxite, Ark. Ten years after a change in the common water supply. U.S. Public Health Reports. 53:1736-1748.

Deninger, A. News and notes. (1907) Dent. Rec. 27:122.

Dickson, S. (1969) Class attitudes to fluoridation. Health Educ. J. 28:139-149.

Docev, N., and Kalajdziev, G. (1973) Use of Ovonsniste fluorine mineral water in drop, capsule and tablet form for caries prophylaxis. Kurortol Fizioter (Sofia) 10:180-181.

Douglas, T.E. (1957) Fluoride dentifrice and stomatitis. Northwest Med. 56:1037-1039.

Dunning, J.M. (1965) Current status of fluoridation. New Eng. J. Med. 272:30-34, 84-88.

Eager, J.M. (1901) Denti di Chaie (Chiaie teeth). Public Health Rep. (Wash.) 16:2576-2577. Idem (1902) Chiaie teeth. Dent. Cosmos. 44:300-301.

Erasquin, R. (1935) Dientes veteados. Rev. Odont. (Buenos Aires) 23:296-313.

Erhardt, A. (1874) Kali fluoratum zur Erhaltung der Zähne. Memorabil. Mschr. ration. Aerzte. 19:359-360.

Ericsson, Y. (1973) Effect of infant diets with widely different fluoride contents on the fluoride concentrations of deciduous teeth. Caries Res. 7:56-62.

Ericsson, Y., and Ribelius, U. (1970) Increased fluoride ingestion by bottle-fed infants and its effect. Acta Paediat. Scand. 59:424-426.

Ericsson, Y., and Ribelius, U. (1971) Wide variations of fluoride supply to infants and its effect. Caries Res. 5:78-88.

Flaumenhaft, E. (1971) Fluoridation of drinking water in The Netherlands. Zahnaerztl. Mitt. 6:886-887.

Forrester, D.J., and Schulz, E.M. (editors) (1974) International Workshop on Fluorides and Dental Caries Reductions. University of Maryland School of Dentistry, Baltimore.

Frankel, J.M., and Alluklan, M. (1973) Sixteen referenda on fluoridation in Massachusetts; an analysis. J. Public Health Dent. 33:96-103.

Furseth, R., and Mjör, I.A. (1973) Pulp studies after 2 percent sodium fluoride treatment of experimentally prepared cavities. Oral Surg., Oral Med., Oral Path. 36:109-114.

Fynn, H.A. (1910) Some remarks on the defects in enamel of the children of Colorado Springs. Items of Interest 32:31-34.

Gabowitsch, R.D., and Owrutzki, G.D. (1969) Fluor in der Stomatologie und Hygiene. Kasan, p. 339, [Cited from Künzel, 1970a].

Gaum, E., Cataldo, E., and Shiere, F. (1973) Reaction of the gingiva to acidulated fluoride gel. J. Dent. Child. 40:22-26.

Gwinnet, A.J. (1973) The sequence of topical fluoride and occlusal sealant applications. J. Amer. Soc. Prevent. Dent. 3:54-57.

Hannon, F. (1926) Letter to the Editor. Water Works Eng. 79:934.

Hars, E., and Massler, M. (1973) Effects of fluorides, corticosteroids and tetracyclines on extraction wound healing in rats. Acta Odontol. Scand. 30:511-522.

Heifetz, S.B., and Horowitz, H.S. (1974) Effect of school water fluoridation on dental caries: interim results in Seagrove, N.C., after four years. J. Amer. Dent. Assoc. 88:352-355.

Helöe, L.A., and Birkeland, J.M. (1974) The public opinion in Norway on water fluoridation. Community Dent. Oral Epedemiol. 2:95-97.

Hempel, W., and Scheffler, W. (1899) Über eine Methode zur Bestimmung des Fluors neben Kohlensäure und den Fluorgehalt von einigen Zähnen. Zeitschrift Anorganische Chem. 20:1-11.

Hennon, D.K.,Stookey, G.K., and Muhler, J.C. (1969) Blood and urinary fluoride levels in humans associated with ingestion of sodium fluoride-containing vitamin tablets. J. Dent. Res. 48:1211-1215.

Hennon, D.K., Stookey, G.K. and Muhler, J.C. (1972) Prophylaxis of dental caries: Relative effectiveness of chewable fluoride preparations with and without added vitamins. J. Pediatr. 80:1018-1021.

Heuser, H., Schmidt, H.F.M., and Krumme, S.T. (1970) The reaction of the healthy tooth pulp to fluorides. Oest Z. Stomat. 67:448-456.

Hodge, H.C., and Finn, S.B. (1939) Reduction in experimental rat caries by fluorine. Proc. Soc. Exper. Biol. Med. 42:318-320.

Hodge, H.C., Luce-Clausen, E.M., and Brown, E.F. (1939) Fluorosis in rats due to contamination with fluorine of commercial casein. The effects of darkness and of controlled radiation upon the pathology of the teeth. J. Nutr. 17:333-346.

Hodge, H.C., and Smith, F.A. (1965) In Fluorine Chemistry (J.H. Simons, editor). Academic Press, New York, Vol. 4.

Hodge, H.C., and Smith, F.A. (1968) Fluorides and man. Ann. Rev. Pharmacol. 8:395-408.

Hodge, H.C., and Sognnaes, R.F. (1946) Experimental caries and the mechanism of caries inhibition by fluorine. In Dental Caries and Fluorine (R.F. Moulton, editor) Amer. Assoc. Adv. Science, Washington, D.C., p.53-73.

Hoppert, C.A., Webber, P.A., and Canniff, T.L. (1931) The production of dental caries in rats fed an adequate diet. Science 74:77.

Horowitz, H.S., Heifetz, S., and Law, F. (1972) Effect of school water fluoridation on dental caries: final results in Elk Lake, Pa. after 12 years. J. Amer. Dent. Assoc. 84:832-838.

Hurny, T. (1974) The fluoride tablets war in Austria. Schweiz. Monatschr. Zahnheilkd. 84:427-431.

Jackson, D. (1972) Attitudes to fluoridation. A survey of British housewives. Brit. Dent. J. 132:219-222.

Jenkins, G.H. (1969) Fluoride and tea. Lancet 2:542.

Jiraskova, M. (1967) Prophylaxis of dental caries in Czechslovakia, its present and future development. Odont. Revy 17:67 (Suppl. No. 10).

Kerley, M.A. (1973) Incisor fluorosis in mice fed fish protein concentrate. J. Dent. Res. 52:626.

King, J.D. (1935) Dietary factors in the production of dental disease in experimental animals, with special reference to the rat. Part I: Dental caries. Brit. Dent. J. 59:233;305-316.

Klein, H., and McCollum, E.V. (1931) A preliminary note on the significance of the phosphorus intake in the diet and blood phosphorus concentration in the experimental production of caries-immunity and caries-susceptibility in the rats. Science 74: 662-664.

Knutson, J.W. (1970) Water fluoridation after 25 years. J. Amer. Dent. Assoc. 80:765-769.

Knutson, J.W., and Amstrong, W.D. (1944) Post-war implications of fluorine and dental health -- the use of topically applied fluorine. Amer. J. Public Health 34:234-239.

Koch, G., and Petersson, L.G.   (1975)   Caries preventive effect of a fluoride-containing varnish (Duraphat) after 1 year's study.   Community Dent. Oral Epidemiol. 3:262-266.

Kondo, T., Yasaki, T., and Okudera, H.   (1973)   The absorption and urinary excretion of fluorine after ingestion of green tea. J. Dent. Health 23:45-51.

Kostlan, J., and Jiraskova, J.   (1967)   Fluorizace vody v Ceskoslovensku. Csl. Stomatologia 59:299.

Kota, K., Katoh, S., and Hosoda, H.   (1973)   Evaluation of dental amalgam containing fluorides. II.   Mode of fluorine penetration into preparations walls.   Jap. J. Conserv. Dent. 16:66-73.

Kridakorn, O.   (1973)   Urinary estimation of optimal fluoride dosage with fish sauce. J. Dent. Assoc. Thai 23:72-82.

Kühns, (1888)   In discussion of Prof. Dr. W.D. Miller's talk on "The effect of nutrition, and especially calcium salts on the teeth."   Proc. 14th Congress, Dental Society of+ lower Sachsen [Hannover, 8 and 9 July, 1888], Dtsch. Mschr. Zahnheilk. 6:446.

Künzel, W.   (1967)   Bericht über ein Internationales Symposium über die Kariesprophylaxe mit Fluoriden (Sofia, Bulgaris, 22-25 Mars, 1967) Dtsch. Stomatolol. 17:636.

Idem   (1970)   Ten years of drinking water fluoridation in the German Democratic Republic. Deutsch Gesundh. 25:1197-1203.

Idem   (1970a)   Present condition and prospects of fluoridation of drinking water in European socialist countries. Nederl. T. Tandheelk. 77:407-410.

Idem   (1974)   Results of fluoridation as a generalized caries-prevention measure from an international viewpoint. Österr. Z. Stomatol. 71:322-333.

Lange, S., Muhlemann, H.R., Hotz, P., and Son, S.   (1971)   Cytological evaluation of chemotoxic effects on the oral mucosa. Helv. Odont. Acta 15:127-129.

Lemke, C.W., Doherty, J.M., and Arra, M.C.   (1970)   Controlled fluoridation: the dental effects of discontinuation in Antigo, Wisconsin. J. Amer. Dent. Assoc. 80:782-786.

Librus, H., Pietrokovski, J., Ulmanski, M., and Gedalia, I.   (1973)   The effect of fluoride on molar socket healing in the rat.   Arch. Oral Biol. 18:1283-1290.

Lilly, C.A.   (1938)   Lessened incidence of caries when casein replaces milk in the coarse corn diet. Proc. Soc. Exper. Biol. Med. 38:398.

Lindhe, J., Hansson, B.O., and Branemark, P.I. (1971) The effect of topical application of fluorides on the gingival tissues. J. Periodont. Res. 6:211-217.

Louridis, O., Theodossiuo, A., Demetriou, N., and Bazopoulou-Kyrkanidou, E. (1971) Influence of absorbed fluorides in gingival inflammation. Relation between calculus, degree of periodontal disease, deep migration of the epithelial attachment and gingival inflammation in white rats. Rev. Stomat. (Paris) 72:297-299.

Ludwig, T.G. (1971) Hastings fluoridation project. VI. Dental effects between 1954 and 1970. New Zeal. Dent. J. 67:155-160.

Lukacs, A. (1972) Preliminary studies and theoretical basis of a new method of mass fluoridation. Rass. Int. Stomat. Prat. 23:125-133.

Maiwald, H.J. (1974) Three years of topical application of a fluorine varnish as a collective prophylactic measure against caries. Stomatologie (DDR) 24:123-125.

Martin, N.D. (1972) Optimal fluoride intake. Med. J. Aust. 1:1118.

Masaki, T. (1931) Geographic distribution of mottled teeth in Japan. Shikwa Gakuho 36:875-897.

McClure, F.J. (1970) Water Fluoridation; the Search and the Victory. U.S. Dept. of HEW. National Intstitutes of Health. National Institute of Dental Research, Bethesda, Md.

McKay, F.S. (in collaboration with Black, G.V.) (1916) An investigation of mottled teeth: An endemic developmental imperfection of the enamel of the teeth heretofore unknown in the literature of dentistry. Dental Cosmos 58:(part I) 477-484, (part II) 627-644, (part III) 781-792, (part IV) 894-904.

McKay, F.S. (1926) Water supplies charged with disfiguring teeth. Water Works J. 79:71-72; 79-80.

Idem (1928) The relation of mottled enamel to caries. J. Amer. Dent. Assoc. 15:1429-1437.

Idem (1929) The establishment of a definite relation between enamel that is defective in its structure, as mottled enamel, and the liability to decay. Dental Cosmos 71:747-755.

Idem (1933) Mottled enamel: The prevention of its further production through a change of water supply at Oakley, Idaho. J. Amer. Dent. Assoc. 20: 1137-1149.

Mejia, R.V., Espinal, F.T., and Velez, H.A. (1972) Fluoridation of salt in four Colombian communities. II. Basic study of dental caries. Bol. Ofic. Sanit. Panamer. 73:561-571.

Michel, A. (1897) Untersuchungen über den Fluorgehalt normaler und cariöser Zähne. Deutsch. Mschr. Zahndeilk. 15:332-338.

Mieler, I., Kittler, G., and Mieler, W. (1972) The behavior of the mitotic coefficient in interdental papillae after application of fluorides. Deutsch. Stomat. 22:841-844.

Miller, B.F. (1938) Inhibition of dental caries in the rat by fluoride and iodoacetic acid. Proc. Soc. Exper. Biol. Med. 39:389-393.

Möller, I.J. (1965) Dental Fluorose og Caries. Rhodos International Science Publ. Copenhagen.

Idem (1969) The systemic administration of fluorides in the prevention of dental caries. Ugeskr. Laeg. 131:2128-2135.

Morichini, D. (1805) Analisi dello smalto di un dente fossile di elefante e dei denti. humani. Mem. Fis. Soc. Ital. Sci. 12:73-88.

Moulton, F.R. (editor) (1942) Fluorine and Dental Health. Amer. Assoc. Adv. Science, Washington, D.C.

Idem (1946) Dental Caries and Fluorine. Amer. Assoc. Adv. Science, Washington, D.C.

Muhler, J.C. (1965) Fifty-two Pearls and their Environment. Indiana University Press, Bloomington, Ind.

Myers, C.L., Long, R.P., Balser, J.D., and Stookey, G.K. (1973) In vivo alterations in enamel from zinc phosphate cement containing stannous fluoride. J. Amer. Dent. Assoc. 87:1216-1222.

Myers, C.L., Stanley, H.R., and Heyde, J.B. (1971) Response of the primate dental pulp to a concentrated stannous fluoride solution. J. Dent. Res. 50:517.

Naumann, E. (1970) Fluoridation of drinking water for prevention of caries, yes or no? Oeff. Gesundheitswesen 32:163-172.

Newbrun, E. (editor) (1975) Fluorides and Dental Caries. Second edition. Charles C. Thomas, Springfield, Ill.

Idem (1977) The safety of water fluoridation. J. Am. Dent. Assoc. 94:301-304.

Petersen, G. (1970) High fluoride content in teeth of Greenlandic school children despite low fluoride content in drinking water. Tandlaegebladet 74:1215-1224.

Petterson, E.O. (1969) The decision to fluoridate: the impact of the elected officials and the community power structure. J. Public Health Dent. 29:153-169.

Idem (1971) Opinions about water fluoridation among dental students. Tandlaek. Tidn. 63:872-877.

Idem (1972) Abolition of the right of local Swedish authorities to fluoridate water. J. Public Health Dent. 32:243-247.

Poulsen, S., and Meller, I.J. (1974) Gingivitis and dental plaque in relation to dental fluorosis in man in Morocco. Arch. Oral Biol. 19:951-954.

Rodriguez, F.E. (1916) The United States Indian Service Field Dental Corps. A new field of activity in which physician, dentist and teacher collaborate. Trans. Panama Pacific Dent. Congress, San Francisco, Cal. (Aug. 30 - Sept. 9, 1915) p. 291-304.

Rosebury, T., Karshan, M., and Foley, G. (1933) Studies in the rat of susceptibility to dental caries. Part III. J. Dent. Res. 13:379-398.

Rybakov, A.A. (1968) Stomatologische Prophylaxe bei Kindern als Grundlage für die Organisation der stomatologischen Betreuung der Bevölkerung, Moscow, (Cited from Künzel, 1970).

Sanchez Y Castillo, J., and Gomez-Castellanos, A. (1973) The problem of fluorosis in Mexico. Assoc. Denta. Mexicana 30:37-39.

San Filippo, F.A., and Battistone, G.C. (1971) The fluoride content of a representative diet of the young adult male. Clin. Chim. Acta 31:453-457.

San Filippo, F.A., Battistone, G.C., and Chandler, D.W. (1972) The fluoride content of Army field rations. Milit. Med. 137:11-12.

Schmidt, H.J. (1972) Milk-fluoridation for the prevention of dental caries. Öst. Z. Stomat. 69:426-432.

Shaw, J.H. (editor) (1954) Fluoridation as a Public Health Measure. Amer. Assoc. Adv. Science, Washington, D.C.

Smith, M.O., Lantz, E.M., and Smith, H.V. (1931) The cause of mottled enamel, a defect of human teeth. Univ. Arizona Techn. Bull. No. 32:253-282.

Sognnaes, R.F. (1940) Fluor -- en faktor i karies profylaksen? I. Oversikt over noen nyere undersökelser med saerlig henblikk pa amerikansk litteratur (Eng. summary). Norske Tannlaegefor. Tid. 50:147-153.

Idem (1941) Effect of topical fluorine application on experimental rat caries. Brit. Dent. J. 70:433-437.

Idem (1941a) A condition suggestive of threshold dental fluorosis observed in Tristan da Cunha. I. Clinical condition of the teeth. J. Dent. Res. 20:303-313.

Idem   (1954)   Oral Health Survey of Tristan da Cunha.
Results of the Norwegian Scientific Expedition to
Tristan da Cunha during 1937-1938, Det Norske
Videnskaps-Akademi, Oslo. Pub. No. 24:1-145.

Idem   (1954a)   Relative merits of various fluoridation
vehicles. In Fluoridation as a Public Health Measure
(J.H. Shaw, editor) Amer. Assoc. Adv. Science,
Washington, D.C., 1954a.

Idem   (1965)   Fluoride protection of bones and teeth.
Some changing concepts of dosage and application are
shedding new light on fluoridation.   Science
150:989-993.

Idem   (Guest editor)   (1967)   Fluorides for better bones
and teeth. Introduction to Orthopaedic Symposium.   In
Clinical Orthopaedics, No. 55:3-4.

Sognnaes, R.F., and Armstrong, W.D.   (1941)   A condition
suggestive of threshold dental fluorosis observed in
Tristan da Cunha. II. Fluorine content of the teeth.
J. Dent. Res. 20:314-322.

Sognnaes, R.F. (Chairman), Arnold, F.A., Jr., Hodge, H.C.,
and Kline, O.L.   (1953) The problem of providing
optimum fluoride intake for prevention of dental
caries. National Acad. Sciences, Natl. Res. Council
Report.   Pub. No. 294:1-15.

Stamm, J.W.   (1972)   Milk fluoridation as a public health
measure. J. Canad. Dent. Assoc. 38:446-448.

Stillings, B.R., Lagally, H.R., Zook, E., and Zipkin, I.
(1973)   Further studies on the availability of the
fluoride in fish protein concentrate. J. Nutr.
103:26-35.

Torell, P., and Siberg, A.   (1962)   Mouthwash with sodium
fluoride and potassium fluoride. Odont. Revy.
13:62-71.

Toth, K.   (1973)   Fluoridation of domestic salt after three
years. Caries Res. 7:269-272.

Tsutsui, M., Utsumi, N., Shimahara, T., Ashida, C., Mizuno,
S., and Ikuma, K.   (1970) Histopathological studies of
pulp tissue to fluoride applied to freshly cut dentin:
effects on SnF^2 and $NaF-H^3PO^4$.   J. Osaka Odont.
Soc. 33:543-544.

Ulvestad, H.   (1973)   The fluorine content of tea. Norske
Tannlaegeforen. Tid. 83:495-497.

Velu, H.   (1931)   Dystrophie dentaire des Mammiferes des
zones phosphatees (darmous) et fluorose chronique.
Compt. Rend. de la Soc. de Biol. 108:750-752.

Volker, J.F.   (1942)   Fluorosis studies at the University
of Rochester.   In Fluorine and Dental Health (F.A.
Moulton, editor) Amer. Assoc. Adv. Science,
Washington, D.C., p. 74-80.

Wigdorowicz-Makowerowa, N.   (1969)   Drinking water
    fluoridation in Poland. Postepy. Hig. Med. Dosw.
    23:879-880.
Wigdorowicz-MaKowerowa, N., Potoczek, S., and Rakowicz, R.
    (1970) The fluoridation of water in Poland and its
    expected results.   Postepy. Hig. Med. Dosw.
    24:291-299.
Williams, J.L. (1923) Mottled enamel and other studies of
    normal and pathological conditions of this tissue. J.
    Dent. Res. 5:117-195.
Wrampelmeyer, E.   (1893)   Über den Fluorgehalt der
    Zähne. Z. Anal. Chem. 32:550-553.
World Health Organization.   Fluorides and human health.
    WHO, Geneva, 1970.
World Health Organization (1975)   WHO launches new
    fluoridation effort. Fluoridation Reporter, No. 13,
    Amer. Dent. Assoc., Chicago, Ill.

Acknowledgments

The author acknowledges with much appreciation and
admiration the assistance of Molly Blaschke, librarian,
Alice Lerch, secretary, and Rhoda Freeman, Supervisor,
Word Processing Center, UCLA School of Dentistry.

2

Fluoride Supplements for Systemic Effects in Caries Prevention

Thomas M. Marthaler

Introduction

The first reports on artificial water fluoridation published in the early fifties stressed the importance of systemic, pre-eruptive fluoride. Twenty years ago, Arnold (1957) recognized the importance of posteruptive effects of waterborne fluoride for the first time. This is well illustrated by the following quotation: "Another observation of special interest in the Grand Rapids study is the result on the older age groups of children. When the study was started it was generally assumed that the beneficial effects of fluoridation, if they did occur, would be observed in persons whose teeth were formed while the individual was using the fluoridated water supply. Little if any benefit was expected on teeth which had already been formed, at least to the extent of calcification of the coronal portion of the teeth. The results of the 16 year olds, for example, do not support this hypothesis. There have been definite and significant reductions in dental caries prevalence in this group of children. It is to be remembered that these children in most cases were those who presumably had the coronal portion of their permanent teeth already calcified when fluoridation started."

The interest of dental investigators and practitioners has since started to focus on posteruptive or topical effects of fluoride, while the systemic effects have been studied on a limited basis only. Attempts to separate the systemic from the topical cariostatic effects have been rare. With respect to water fluoridation this has not been important since this mass method of prevention provides both systemic and topical fluoride.

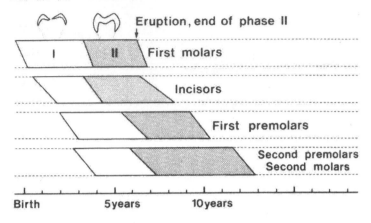

Phase I: Mineralisation of crown
Phase II: Unerupted, fully mineralized crown

Figure 1.   Phase I is the period of primary mineralization.
Phase II begins when the crown is mineralized, but not yet
erupted.   Systemic fluoride reaches the crown during Phases
I and II.   Premolars are also called bicuspids.

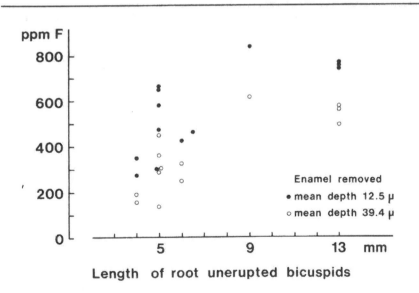

Figure 2.   Fluoride concentration in enamel according to the
length of the root.   All teeth were bicuspids extracted for
orthodontic reasons prior to eruption.   Cumulative concentrations for enamel layer thickness 0 to 12.5 μ (standard deviation 2.4 μ) and 0 to 39.4 μ (standard deviation 6.1 μ).
Data from Mühlemann (1965).

The distinction between the systemic and topical benefit is important with respect to the fluoridation of domestic salt and the use of fluoride tablets. Accordingly, at the "International Workshop on Fluorides and Dental Caries Reduction," held in Baltimore in 1974, the committee reporting on fluoride tablets suggested further investigations "to determine the benefit of fluoride tablets administered during calcification periods" (Forrester and Schulz, 1974, p. 110). Nevertheless, the committee concluded that "a principal benefit of fluoride supplements, which reduce caries, appears to be topical (post-eruptive)" (p. 111). Detailed knowledge regarding systemic and topical cariostatic effects is also important with respect to fluoridation of domestic salt. Salt intake is very low during the first and second years of life when the crowns of deciduous molars and the first permanent molars calcify.

Systemic fluoride reaches the tooth at two different developmental stages. During phase I enamel and dentin of the crown are in the process of mineralization, while during phase II the crown is fully formed but is unerupted. The approximate duration of phase I and phase II is given in Fig. 1. For permanent teeth, each phase lasts three to six years.

The complicated processes involved in the formation of enamel and dentin (phase I) have been the object of numerous studies. Once enamel is fully formed at the end of phase I, enamel fluoride concentrations at depths of 50 μ or more below the surface will remain fairly constant. These concentrations normally vary between 50 and 150 ppm according to the supply of fluoride during crown formation. Fluoride levels in primary crown dentin, formed at the same time as enamel, are also largely determined by the fluoride supply to the body during its formation. Phase I is considered to end when the enamel appears as fully mineralized on ordinary dental radiographs.

Fluoride levels in enamel and dentin are compiled in Table 1. In deep enamel, fluoride levels were at 36 and 42 ppm when the fluoride content of the drinking water was below 0.4 ppm. Higher concentrations, ranging from 48 to 113, were found when the drinking water contained 0.4 to 1.0 ppm. Levels of 129 and 245 ppm were associated with water fluoride content of 2.1 to 3.0 ppm. Dentinal fluoride concentrations were higher, but similarly associated with fluoride content of the drinking water.

During phase II the crown moves very slowly toward the gingivae. The tooth is bathed in interstitial fluid for three to six years, and the surface enamel acquires fluoride

Table 1. Fluoride in Deep Enamel* and in Dentin

Fluoride history	Age at extraction	Fluoride (µg/g) Enamel	Dentin	Ref.[†]
Water (ppm F)				
0.1	20	42	–	1
0.2	10-11	36	–	2
0.4-1.0	25	113	–	3
1.0	20	48	–	4
	20-29	105	–	
	20-29	98	220-301	5
2.1	20	129		1
2.9	20	245		5
3.0	20	152	720-1268	5
No fluoride supplements	10-12	28	79	6
F-tablets irregularly	10-12	43	110	
F-tablets regularly	10-12	46	155	

* Depth of surface enamel removed, 50-200 µ.

[†] References: 1) Isaac et al. (1958); 2) Koch and Friberger (1971; 3) Nicholson and Mellberg (1969); 4) Brudevold et al. (1956); 5) Yoon et al. (1960); 6) Baumgartner et al. (1976).

as evidenced by the study of Brudevold (1962) and illustrated by the data shown in Fig. 2. This is part of the
process referred to as pre-eruptive enamel maturation. It
should be kept in mind that after mineralization further
changes in the enamel are governed by the physico-chemical
processes such as diffusion, osmosis, and solid-solution
exchange reactions. Dentin is continually formed and incor-
porates fluoride according to the intake levels. Phase II
ends with the eruption of the tooth.

Phase III denotes the post-eruptive life of the tooth.
Systemic effects are now limited to cementum and dentin,
and especially to the layers adjacent to, or newly formed by,
the dental pulp. The influence of saliva may be regarded as
systemic, but it is preferable to view it as a local effect.
Whole saliva, or actual oral fluid, is a mixture of saliva
secreted by the various types of glands. It contains epithelial and bacterial cellular elements and fragments, as
well as substances brought into the mouth from the outside.
Salivary fluoride levels are as low as 0.01 to 0.03 ppm.
Büttner et al. (1973) found that levels of ionized salivary
fluoride are consistently at one-fourth of total plasma
fluoride concentration. This ratio was maintained during
experiments in which large doses of fluoride were used to
raise plasma fluoride, normally at 0.02 to 0.05, up to
1.4 ppm.

At the start of phase III, i.e., shortly after eruption,
root dentin and some additional coronal dentin is formed.
Dentinal fluoride levels continue to increase substantially
up to the age of 50 (Jackson and Weidmann, 1959), but the
increase in fluoride concentration in the total mass of
dentin is largely due to higher levels of fluoride in later
formed dentin. Yoon et al. (1960) found that dentin adjacent to the pulp and the root surface (cementum and some
dentin) contains much more fluoride than primary dentin.
This difference becomes more pronounced with increasing age.

Systemic effects of water fluoridation

The systemic effect of fluoride is most apparent in
water fluoridation studies. Selected data of children aged
9 and 13 years reported in the Grand Rapids Study are shown
in Table 2.

In children aged 9 years, the number of DMF* teeth had
fallen from the initial average of 3.90 to 3.12 by 1947.

---

* Upper case letters used for permanent teeth, lower
case for deciduous teeth. D, decayed; M, missing; F, filled;
T, teeth; E, extracted; S, surfaces.

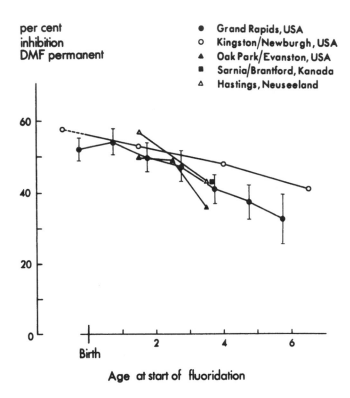

Figure 3. Percent inhibition of caries in the permanent dentition (reductions of DMF tooth counts) according to the age at which the children were first exposed to fluoridated water. (For detailed tables refer to Marthaler, 1968.)

The latter figure relates to children who had been exposed to fluoridated water since the age of 6 years. However, in 1952 an average of only 2.02 DMF teeth was found in 9 year old children with fluoride exposure since birth. Pre-eruptive fluoridation had provided the greater part of the benefit. The great majority of teeth becoming carious, i.e., decayed, missing or filled or DMF prior to age 9 are first molars which erupt at age 6.

Table 2. Decayed, Missing, and Filled Teeth (DMFT) as a Function of Exposure to Fluoridated Water*

| | Years of exposure prior to examination | | | |
	0	3	8	13
Nine year olds	3.90	3.12	2.02	--
Thirteen year olds	9.73	8.47	5.87	3.86

* Values are average DMFT per child. Data are compiled from Arnold et al. (1956, 1962).

Children of age 13 had 9.73 DMF teeth in 1944-45. This mean had fallen to 5.87 in children exposed to fluoride since the age of 5 (at the 1952 examinations). Children examined in 1957, who were born at about the time of commencement of fluoridation (March 1945) showed only 3.86 DMF teeth. In this age group, two teeth on average (5.87-3.86) were saved from becoming DMF when water fluoridation was started at birth instead of at age 5.

The summarized findings of several water fluoridation studies are given in Figure 3. For the Grand Rapids Study with examinations every year, it was possible to compute confidence limits (p = 0.95) based on several age groups with fluoride supplementation from various ages, 0 to 6 (Marthaler, 1968). The results of all studies indicate that there is little difference in the percentage reduction (50% to 60%) whether fluoride is given from birth or from the age of one or two years. Moreover, all studies indicate a reduction of less than 50% when water fluoridation is begun at age four or later.

Connor and Harwood (1963) gave separate data for first molars and remaining teeth in children 12 to 14 years of age from Brandon (Manitoba, Canada). In this city, water fluoridation was introduced in 1955. The number of DMF first molars was almost identical in 1955 (3.20), 1960 (3.33), and 1962 (3.13). However, among the remaining teeth there were 4.19 DMF in 1955 but only 2.47 in 1960. The data are shown in more detail in Table 3. The results of the study indicated that the only permanent tooth requiring fluoride at a very early age is the first molar.

Table 3. DMFT Scores for First Molars and Other Teeth in Children Aged 12 to 14*

	Years of exposure		
	0	5	7
First molar	3.20	3.33	3.13
Other teeth	4.19	2.47	1.64

* Data from Brandon, Manitoba, Canada (Connor and Harwood, 1963).

However, the practical importance of systemic, pre-eruptive fluoride cannot be inferred from tooth counts but should be based on analysis of different sites of caries predilection within the teeth. In first and second molars it is fissure caries and, to a lesser degree, caries of pits that mainly determine whether these teeth are DMF, since the remaining surfaces are less susceptible to caries. This has been most clearly demonstrated by Toverud et al. (1961).

Russell and Hamilton (1961) have provided pertinent first-molar data allowing distinction between pit and fissure caries on the one hand and free smooth surface caries on the other (Table 4). Upper lingual and lower buccal data were pooled since DMF experience of these surfaces is largely determined by caries of pits and grooves on both of those surfaces. In addition, caries experience may be combined with occlusal caries, which is always of the fissure type. Conversely, upper buccal and lower lingual surfaces are essentially free smooth surfaces. Up to age 12, the distal surface is also a free smooth surface. With the eruption of the second molar, it becomes an approximal smooth surface. The mesial surface was omitted since decay at this site is often locally induced by the presence of a carious surface of the second deciduous molar.

In children 9 to 11 years of age, the reduction of
DMFS experience in occlusal and pitted buccolingual, i.e.,
nonsmooth,surfaces was 40% while for smooth surfaces it
reached 71%. These chilren were 1, 2, or 3 years old when
water fluoridation was initiated in 1952. Children of the
age group 12 to 14 had been first exposed to fluoride at
ages 4, 5, and 6. DMF experience on nonsmooth surfaces was
decreased by only 9 percent, whereas smooth surface DMF
counts were reduced by 34%. It is apparent that early administration of fluoride is most important for the prevention of fissure caries.

No detailed data for other types of teeth are available. Kwant et al. (1973) presented separate data for pit
and fissure, approximal, and free smooth surface caries in
children with fluoride from birth, from age 4, and from
age 9. The average results in children 13 and 15 years of
age are given in Table 5. There was a 14% reduction of pit
and fissure caries when fluoridation was started at age 8 or
9. The reduction reached 29% when it was started at age 4,
and 39% when it was started at birth. For approximal surfaces, the reduction was already at 65% when fluoridation
was started at age 4 while the maximum reduction, obtained
by fluoridation since birth, was 75%. As to free smooth
surface caries-predilection sites, reduction was 65% even
when fluoridation was started at age 8, at a time when all
crowns were fully mineralized. These findings support two
hypotheses formulated by Backer-Dirks (1963): "(1) Fissures
will always benefit less than approximal surfaces from water
fluoridation, and (2) in order to have an important effect
on fissure caries, the extra fluoride must be present in an
early stage of tooth formation." It should be noted that
during the sugar restriction period of World War II, the
smallest benefit was also seen in fissures (Toverud et al.,
1961).

From all the data presented, it is evident that the
systemic fluoride effect is most pronounced with respect to
prevention of fissure caries. A clear-cut systemic effect
is obtained during phase II, i.e., during the postdevelop-
mental, pre-eruptive period, when the fully formed enamel
acquires fluoride from the interstitial liquid. Regarding
free smooth surfaces, the greater part of the protection
provided by systemic fluoride is obtained during phase II.

The risk of caries begins with the eruption of the
predilection site. However, the first symptoms of the
clinical caries in fissures and pits become apparent soon
after eruption, whereas clinical caries of free smooth surfaces usually develop only several years after eruption.
This may explain why the benefit provided to pits and

Table 4. Expected and Observed DMFS of First Molars*

	Occlusal fissure	Pitted[†] surface	Total nonsmooth	Smooth[†] surface	Distal surface	Total smooth
Age 9-11, 696 children, fluoride since age 1-3						
Expected	2.36	0.88	3.24	0.50	0.42	0.92
Observed	1.43	0.52	1.95	0.15	0.12	0.27
Percent change	- 39	- 41	- 40	- 70	- 71	- 71
Age 12-14, 1358 children, fluoride since age 4-6						
Expected	2.96	1.36	4.32	0.82	0.92	1.74
Observed	2.72	1.19	3.91	0.58	0.57	1.15
Percent change	- 8	- 12	- 9	- 29	- 38	- 34
Age 15, 140 children, fluoride since age 7						
Expected	3.03	1.49	4.52	1.01	1.03	2.04
Observed	3.27	1.77	5.04	1.24	1.22	2.46
Percent change	+ 8	+ 19	+ 12	+ 23	+ 18	+ 21

* Data from Russell and Hamilton (1961). Expected values are the DMFS that would have occurred if the 1952 attack rates had persisted (water fluoridation was introduced in 1952); observed values are based on examinations made in 1960.

[†] Pits and grooves generally occur on the lingual surfaces in upper molars and on the buccal surfaces in lower molars.

fissures depends on the pre-eruptive fluoride intake. Regarding smooth surfaces, the first post-eruptive years without clinical signs of caries may allow accumulation of sufficient amounts of fluoride in surface layers of enamel to warrant a topical preventive effect.

Table 5.   Caries Lesions by Site in 13 and 15 Year Olds as a Function of Exposure to Fluoridated water: Scores and Percent Difference from Controls*

	1944-45**		1949		1953	
	Cul.	Tiel	Cul.	Tiel	Cul.	Tiel
Age at start of F		9/8		4		0
Pit and fissure sites	$12.4^{†}$	10.7	13.6	9.7	13.4	8.2
% Difference		14		29		39
Proximal surfaces	6.2	5.2	6.2	2.2	7.1	1.8
% Difference		16		65		75
Free smooth sites	1.39	0.48	1.90	0.46	2.55	0.32
% Difference		65		76		87

* Compiled from Kwant et al. (1973). Culemborg (Cul.) was the control town; in Tiel, drinking water fluoridation was started in 1953.

** Year of birth.

† Means of data reported by children at age 13 and 15.

With water fluoridation, the increase in fluoride concentration of the oral fluid is small when compared to topical fluoridation methods. Therefore, more time is needed to raise fluoride levels in the surface layers of enamel. This question remains open, however, whether fluoride must be given at a very early age in order to reach the tooth during the mineralization of enamel (phase I) or simply in order to secure a sufficiently long pre-eruptive uptake of fluoride by surface enamel (phase II).

In the deciduous dentition, there is only a short interval between termination of crown mineralization and tooth eruption. Crowns of first deciduous molars are fully mineralized at age 6 months, and they erupt at 14 to 18 months. Crowns of second deciduous molars are fully mineralized at 12 months, and eruption takes place at 24 to 30 months.

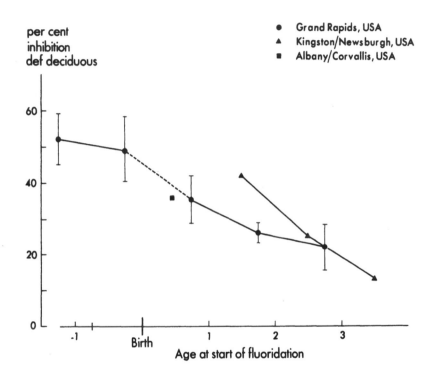

Figure 4.   Percent inhibition of caries in the deciduous dentition (reduction of def tooth counts) according to the age at which the children were first exposed to fluoride water. (For details tables refer to Marthaler, 1968.)

Table 6 shows caries experience in the deciduous dentition related to exposure to fluoridated water. Substantial
protection was provided to the children who received fluoridated water from age 1 to 12 months, which is approximately
16 months before eruption of the deciduous molars. Children
at age 7 years who had been exposed to waterborne fluoride
from age 13 to 24 months had received some benefit. Loss of
deciduous molars was substantially reduced by fluoridation.
Data from similar studies are compiled in Figure 4. It is
evident that fluoridation should start at birth to provide
optimal protection to the deciduous teeth. When exposure
starts at one year of age there is still substantial protection, but little benefit can be expected for the decidu-
oud teeth of children starting to drink fluoridated water
at the age of two or three years.

Table 6. Effect of Fluoridated Water Started in the
First, Second or Third Year of Life on Caries
in Deciduous Teeth*

	1967 Pre-fluoride		1973, after 6 years of fluoridation	
Age 6		--	F since 1-12 months	
df teeth	6.25	(101)	3.74	(129)
Missing molars	1.38		0.23	
Age 7		--	F since 13-24 months	
df teeth	5.36	(60)	4.53	(123)
Missing molars	1.86		0.93	
Age 8		--	F since 25-36 months	
df teeth	4.38	(57)	4.12	(115)
Missing molars	2.92		1.56	

* Data from Medcalf (1975). Numbers of children
examined are given in parentheses. df = decayed, filled.

Data collected in Sweden indicate that wide variations
in fluoride intake during infancy have a comparatively small
influence on fluoride levels in dentin and on later caries
experience of deciduous teeth (Ericsson, 1973; Forsman and
Ericsson, 1974; Forsman, 1974).

Studies with fluoride tablets and school water fluoridation

Investigation of the systemic beneficial effect of fluoride tablets meets considerable difficulty. The teeth must be exposed to systemic fluoride for several years; subsequently, caries can be studied with sufficient reliability only when the teeth have been exposed to the oral environment for several years. It is very difficult to conduct such long-term studies, especially when variations in eruption times are considered (see Fig. 1).

With fluoride tablet studies, there is the additional difficulty of organizing and checking the daily distribution of tablets. Reports of intake at home are not reliable since parents and children tend to confirm that pattern of tablet intake which they feel is expected of them. Accordingly, there are only a few studies providing data relevant to the present topic.

Andersson and Grahnén (1976) studied caries in children who had been given fluoride tablets during the first five years of life (Table 7). According to the authors the results support the hypothesis that "an endogenic cariostatic effect due to fluoride tablets appears to add to a local effect of the mouth rinsing with a fluoridated solution from

---

Table 7. Effect of Regular Use of Fluoride Tablets Until the Age of Five Years on DEF Surfaces*

	Permanent teeth DF surfaces**			Deciduous teeth def surfaces		
	Proximal	Other	Total	Proximal	Other	Total
Control	0.55	3.34	3.89	4.86	3.64	8.50
Fluoride	0.26	2.08	2.34	3.33	2.50	5.83

* Data from Andersson and Grahnén (1976). Caries experiences in 8- to 10-year-old children who consumed fluoride tablets regularly until the age of 5 years and control children from the same school classes. DF = decayed, filled; def = decayed, extracted, filled.

** No permanent teeth extracted.

the age of six." (Fortnightly rinsing with fluoride solutions is a standard practice in Swedish schools.)

In a survey carried out in Switzerland, tablet distribution took place at school some 200 times a year from age
6 or 7 to 14. Cooperation of the teachers was better in the
lower classes than in the upper classes. Under this regimen,
first molars received only post-eruptive fluoride. Second
molars with a similar anatomy received systemic fluoride,
taken up mainly during phase II, but did not receive post-
eruptive fluoride as regularly as did first molars. Since
first molars erupt at age 6 and second molars at approximately age 12, caries experience of first molars at ages 7,
8 and 9 is directly comparable to caries experience of second molars at ages 13, 14 and 15.

Table 8 shows DMF data of first and second molars at
equal post-eruptive ages. Contrary to what might be
expected, first molars consistently showed greater reductions
of DMFS experience, ranging from 39 to 66 percent. Thus,
fluoride tablets distributed fairly regularly during the
first school years were more effective topically in first
molars than systemically in second molars. The results confirmed the reports from teachers, parents, and school children that the distribution was relatively reliable in the
young children but not in the older children. These findings suggest that in the case of fluoride tablets the
topical effect is more important than the systemic effect.
According to Hotz (1969) the average concentration of fluoride in oral fluid remains above 100 ppm for four minutes
during dissolution of 1-mg fluoride tablets. Based on this
finding, a strong topical effect of fluoride tablets may be
assumed. The extremely low caries prevalence in children
with regular tablet intake, as reported by Aasenden and
Peebles (1974), also supports the hypothesis that tablets
provide a strong topical benefit.

School water fluoridation studies may also serve to
distinguish between systemic and topical fluoride effects.
Data from early and late erupting teeth were tabulated separately by Horowitz et al. (1972). Late erupting teeth
experienced a greater benefit than did early erupting ones
(Fig. 5).

Considering the timing of enamel mineralization (Fig. 1)
the additional systemic benefit was obtained principally
during phase II, i.e., fluoride accumulated mainly in the
superficial enamel.

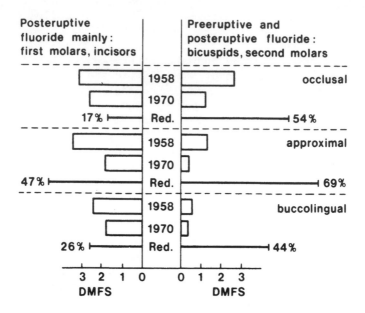

Figure 5. Average number of DMF occlusal, approximal (mesio-distal) and buccolingual surfaces before and after 12 years of school water fluoridation, starting at age 6. Crowns of "early erupting" teeth mineralized during age 0 to 4, but erupt at ages 6 to 8. Crowns of "late erupting" teeth are fully mineralized during ages 2 to 7, but erupt between ages 10 and 12.

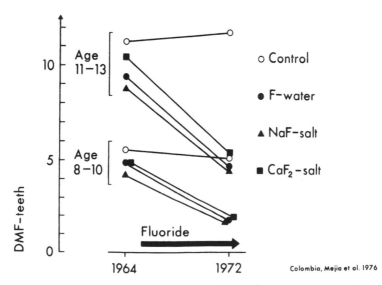

Figure 6. Average number of DMF teeth in 1964 and 1972 of school children in the control and the fluoridated towns.

Table 8. Effect of Fluoride Tablets on Caries
Experience of First Molars at Ages 7 to 9
and Second Molars at Ages 13 to 15*

	Occlusal	Pits and grooves[†]	Free smooth	Mesial
First molars	1.64	0.64	0.31	0.23
% Difference	39	46	67	71
Second molars	2.23	0.28	0.23	0.29
% Difference	21	39	59	66

* Data from Marthaler (1969). Examination of children
in schools with NaF (0.5-1.0 mg) tablets distributed about
200 times a year for 8 years or more. Control children were
examined at the same time and place without the examiner
knowing whether the children were receiving fluoride. Percent difference is based on the controls.

[†] Pits and grooves generally occur on the lingual surfaces in upper molars and on the buccal surfaces in lower
molars.

Alternative means of providing systemic fluoride

Fluoridation of salt (Marthaler et al., 1978; Mejia
et al., 1976; Toth, 1976) and milk has also been investigated. Figure 6 shows caries reductions obtained with NaF
and CaF_2 added to domestic salt, and with fluoride added to
drinking water. Similar caries reductions were obtained
with the three methods. Even in the group of older children
in which part of the cariostatic effect was mediated topically, CaF_2 was as effective as NaF despite its lower solubility. However, salted food does not contain more than
1 or 2 percent NaCl, resulting in fluoride concentrations of
2 to 4 ppm. If there is no excess calcium in the salted
food, fluoride remains in solution and is available for
direct uptake by enamel and plaque.

Fluoridation of milk has been tested clinically in two
studies. Both Rusoff et al. (1962) and Ziegler (1964) found
a cariostatic effect provided by milk supplemented with
fluoride. In view of the short duration of the studies it
must be assumed that the cariostatic effect has been primarily topical.

Table 9.   Effect of Discontinuation of Water Fluoridation
on Caries Experience*

	No. of children	Percent caries-free	DMF teeth	Increase
Second grade				
Dec. 1960	143	71	0.6	
May 1966	94	23	2.0	1.4
Fourth grade				
Dec. 1960	137	35	1.7	
May 1966	86	12	2.9	1.2
Sixth grade				
Dec. 1960	123	26	2.4	
May 1966	85	9	4.6	2.2

* Data from Lemke et al. (1970). Antigo, Wisconsin, DMF-experience in December 1960, when fluoridation was discontinued, and in May 1966. The total number of missing teeth was only 8.

Table 10.   Effect of Length of Time of Fluoride Tablet Use
on Caries Experience*

Age	Tablet use					
	None	Occasional	Daily for indicated no. of years			
			0-2	2-3	3-4	4-5
4-5	2.12	1.72	2.17	1.56	1.17	0.67
5-6	2.75	2.85	2.77	2.59	1.44	0.86

* Data from Fanning et al., 1975. Average number of DMF deciduous teeth per child according to the length of time fluoride tablets were taken.

Does the systemic cariostatic effect persist?

Deatherage (1943) first suggested that protection may diminish or disappear when fluoride supplementation is discontinued. In military personnel who had always lived in areas with optimal fluoride levels in the drinking water he found 6.2 DMF teeth; those who had left the fluoride region some time after the age of 8 showed 7.3 DMF teeth. By 1968, four studies were available confirming an increase of caries activity after discontinuation of supplemental fluoride; three of these studies were made in the U.S. and one in Hungary (Marthaler, 1968).

In 1970, Lemke et al. made similar observations in Antigo, where water fluoridation ended in 1960, 11 years after its inception. The increase in the number of DMF teeth from 1960 to 1966 is displayed in Table 9. Second-grade children showed 0.6 DMF teeth in 1960 but 2.0 in 1966. Their first molars, which largely determine the DMF experience at this age, benefited from systemic fluoride during enamel mineralization of phase I. The lack of fluoride supplementation in later years jeopardized the systemic effect. Similar results were found in fourth-grade children, who showed 1.7 DMF teeth in 1960 and 2.9 in 1966. Fluoride protection was rapidly lost after fluoridation was discontinued.

Discontinuation of daily intake of fluoride tablets is a serious problem. In fact it is the main reason for the poor success rate of such programs. Fanning et al. (1975) collected data on children who had started using tablets at birth but who had later discontinued this measure. Use of tablets up to the age of two did not lead to decreased caries levels (Table 10). However, children who had taken tablets for four years or longer showed a more pronounced difference from the controls than did those who had benefited from water fluoridation. This excellent result may be due to a positive bias. Higher levels of oral hygiene or reduced intake of between-meal sugary snacks were considered to have caused part of the difference.

A detailed analysis of the persistence of the cariostatic effect of topical fluoride showed that several factors must be considered (Marthaler, 1971). Fluoride applied by supervised brushing with plaque removal provided continued protection during a two-year post-treatment period, while fluoride rinses produced no residual effect. Toothbrushing facilitates access of saliva to the enamel surface. The combined action of oral fluid and temporarily high fluoride levels on brushed surfaces may have favored the phenomenon termed posteruptive enamel maturation. Enhanced enamel maturation would probably provide a prolonged or persistently increased relative resistance against caries.

Cariostatic mechanism of fluoride

An international workshop on cariostatic mechanisms of
fluorides held in 1976 (Brown and König, 1977) distinguished
between physico-chemical aspects within the enamel on the
one hand and biochemistry and microbiology of plaque on the
other. It is evident that systemic fluoride accumulated in
the enamel before eruption induces favorable changes in the
physico-chemical properties of enamel. If fluoride acquired
pre-eruptively served as a reservoir and was released to the
plaque, the outer layers of enamel would be rapidly depleted
of fluoride. However, many works have shown that after
eruption additional fluoride is acquired by the enamel, a
process which is often masked by wear or abrasion (Weatherell
et al., 1977).

In sub-surface and especially in the superficial enamel,
there is a gradual increase in fluoride concentration during
phase II. The concentrations in outer enamel from optimally
fluoridated regions exceed 1000 ppm, which, according to
Moreno et al. (1977), does not decrease enamel solubility
sufficiently to protect enamel against the acids formed in
dental plaque. Therefore the remineralization phenomenon
(summarized by Silverstone, 1977) may play an important role.

Apart from these "classical" phenomena, there are other
possibilities for fluoride to inhibit apatite dissolution.
Van der Lugt et al. (1970) has suggested that fluoride in
the apatite crystal has a unique position to block diffusion. Small amounts of fluoride would therefore be able to
reduce the development of caries. Garnier et al. (1976)
studied the kinetics of the dissolution of synthetic apatite
and human enamel. The more fluoride there was in the acidic
buffer, the lower was the rate of dissolution. Brown et al.
(1977) discussed several mechanisms by which fluoride may
render enamel less susceptible to carious demineralization.

Several other mechanisms for the cariostatic effects
of fluoride have been suggested. Tooth crown morphology was
thought to be influenced by fluoride. Findings of altered
crown morphology or tooth size have been reported by some
authors, but not by others (Ericsson, 1977; Grahnén et al.,
1975). If fluoride renders teeth less susceptible to caries,
e.g. by inducing shallower fissure formation, this effect
should be consistently apparent and detected regularly with
different methods. The morphological changes would in addition provide persistent protection irrespective of discontinuation of fluoride. Therefore the marked reductions in
DMF levels consistently observed are not likely to be
explained by changes of morphology.

Initial carious lesions of the enamel such as white spots have a higher fluoride content (Weatherell et al., 1977) than adjacent sound enamel. This shows that once there is carious demineralization, fluoride can penetrate into deeper enamel beyond the outer layers of 20 to 100 µ to which fluoride uptake from topical treatment is limited in general. Fluoride is then undoubtedly trapped by the various types of crystals occurring in partially demineralized enamel and is likely to exert a cariostatic effect in the depth of the lesion. This hypothesis is supported by a study by Hollender and Koch (1969), who found that topical fluoride inhibited progression of radiographic enamel lesions to some extent. Carious areas which are visible on ordinary radiographs must already have undergone considerable demineralization, facilitating diffusion of fluoride and also other ions or molecules.

In sound deep enamel and primary dentin, fluoride levels can only be increased by systemic fluoride during phase I. Elevated fluoride levels in sound deep enamel and in dentin might inhibit the earliest stages of carious demineralization at a time when diffusion of fluoride ions from superficial enamel is still at a minimum. The question of a purely systemic effect of fluoride may be studied in the light of the undermining character of dental decay. As soon as carious demineralization reaches the dentin, it starts to spread towards the dental pulp as well as laterally along the dentino-enamel junction. If fluoride inhibits disintegration of surface enamel but fails to protect the underlying hard tissues, one would expect extensive destruction, frequently not visible at routine dental check-ups, and early loss of teeth. Data on tooth loss can therefore be used to study this question. An old but very thorough statistic of tooth loss is reproduced in Table 11. At ages 25 to 35, mean numbers of teeth lost due to caries in Boulder (low fluoride drinking water) were slightly above two, while less than one tooth was lost due to caries in Colorado Springs (drinking water with 2.5 ppm fluoride). In the age bracket 35 to 44 tooth loss was several times lower in Colorado Springs than in Boulder. In these age groups, 130 examinees had lost 222 teeth because of caries in Colorado Springs while in Boulder 34 persons had lost 337 teeth for this reason. Tooth loss due to periodontitis and other reasons was less frequent. Accordingly, the differences between the two cities are similarly reflected when tooth loss due to any reason is considered. These findings indicate that systemic fluoride available during the period of dentin mineralization retards the carious destruction of dentin.

Delayed eruption is another phenomenon sometimes related to lower DMF levels of children from regions with optimal fluoride levels in the drinking water. As early as

1944, Short et al. reported a delay in eruption when the water contained fluoride at levels above 2 ppm. With 1.0 ppm water a slightly later eruption of permanent teeth occurs in children with formerly high caries levels. The term "delayed eruption" is misleading. The postponement normalizes the eruption times which were changed by premature loss of deciduous teeth. Slightly later loss of deciduous teeth and later eruption of permanent teeth has been observed in a preventive program without systemic fluoride (Marthaler, 1975). As in the water fluoridation studies, the postponement was in the order of a few months only. In many water fluoridation studies, however, eruption times did not appear to change after fluoridation of the water (Ericsson, 1977; Bauer et al., 1974).

Table 11. Comparison of Number of Missing Teeth per Person between Boulder (0.1 ppm F) and Colorado Springs (2.5 ppm F)*

| Age group | No. of Persons Examined | | Cause of Loss | | | |
| | | | Caries | | Other | |
	Bldr.	C.Spr.	Bldr.	C.Spr.	Bldr.	C.Spr.
25–29	41	101	2.15	0.73	0.39	0.16
30–34	29	82	2.03	0.92	1.96	0.22
35–39	22	75	10.05	1.63	1.32	1.58
40–44	12	55	9.67	1.82	2.42	1.27

* From Russell and Elvove (1951)

Concluding comments

During the last two decades interest has shifted from systemic to topical effects of fluoride. The vast majority of recent clinical trials were designed to test topical fluoride effects. Thus, the cariostatic effectiveness of many methods of topical fluoridation has been assessed with remarkable precision. Most experiments lasted from 2 to 4 years and were conducted according to the rules of modern statistical design. The advantage of topical fluoride studies is the ease with which they can be performed.

On the other hand, pre-eruptive or systemic effects of fluoride can only be assessed by studies extending over many years. This may explain why only limited progress has been made in this field. The present analysis indicates that

systemic fluorides are important for obtaining maximum
benefit regarding pit and fissure caries and, to a lesser
degree, regarding approximal surfaces. However, continued
post-eruptive supply of fluorides is indispensable since
otherwise the advantage of systemic fluoride will be lost.

References

Aasenden, R. and Peebles, T. C. (1974) Effects of fluoride
    supplementation from birth on human deciduous and perma-
    nent teeth. Arch. Oral Biol. 19; 321-326.

Andersson, R., and Grahnén, H. (1976) Fluoride tablets in
    pre-school age--effect on primary and permanent teeth.
    Swed. Dent. J. 69; 137-143.

Arnold, F. A., Jr. (1957) The use of fluoride compounds for
    the prevention of dental caries. Int. Dent. J. 7; 54-72.

Arnold, F. A., Jr. et al. (1956) Effect of fluoridated pub-
    lic water supplies on dental caries prevalence. 10th year
    of the Grand Rapids-Muskegon study. Pub. Health Rep. 71;
    652-658.

Arnold, F. A., Jr. et al. (1962) Fifteenth year of the
    Grand Rapids fluoridation study. J. Am. Dent. Assoc. 65;
    780-785.

Backer-Dirks, O. (1963) The assessment of fluoridation as a
    preventive measure in relation to dental caries. Brit.
    Dent. J. 114; 211-216.

Bauer, P. et al. (1974) Eruption bleibender Zähne in
    Gebieten mit niederem und hohem Fluoridgehalt des Trink-
    wassers. Oest. Z. Stomatol. 71; 122-137 and 162-174.

Baumgartner, W. (1976) Der Fluorspiegel in Schmelz und
    Dentin im Lichte anamnestischer Angaben. Med. Diss.,
    Zürich.

Brown, W. E., and König, K. G., Eds. (1977) Cariostatic
    Mechanisms of Fluorides. Proceedings of a Workshop
    Organized by the American Dental Association Health Founda-
    tion and the National Institute of Dental Research, Naples,
    Fla. 1976. Caries Res. 11 (Suppl. 1), 1-327. S. Karger,
    Basel.

Brown, W. E., Gregory, T. M., and Chow, L. C. (1977) Effects
    of fluoride on enamel solubility and cariostasis. In:
    Cariostatic Mechanisms of Fluorides (W. E. Brown and K. G.
    König, eds.). Caries Res. 11 (Suppl. 1), 118-141.

Brudevold, F., Gardner, D. E., and Smith, F. A. (1956) The
    distribution of fluoride in human enamel. J. Dent. Res.
    35; 420-429.

Brudevold, F. (1962) Chemical composition of the teeth in relation to caries. In: Chemistry and Prevention of Dental Caries (R. F. Sognnaes, ed.). Charles C Thomas, Springfield, Ill.

Büttner, W., Henschler, D., and Platz, J. (1973) Karies-prophylaxe durch Fluorid-Einnahme. Dtsch. med. Wschr. 98; 751-756.

Connor, R. A., and Harwood, W. R. (1963) Dental effects of water fluoridation in Brandon, Manitoba: second report. J. Can. Dent. Assoc. 29; 716-777.

Deatherage, C. F. (1943) Fluoride domestic water and dental caries experience in 2026 white Illinois selective service men. J. Dent. Res. 22; 129-137.

Deatherage, C. F. (1943) A study of fluoride domester waters and dental experience in 263 white Illinois selective service men living in fluoride areas following the period of calcification of the permanent teeth. J. Dent. Res. 22; 173-180.

Ericsson, Y. (1973) Effect of infant diet with widely different fluoride contents on the fluoride concentrations of deciduous teeth. Caries Res. 7; 56-62.

Ericsson, Y. (1974) Report on the safety of drinking water fluoridation. Caries Res. 8 (Suppl. 1); 16-27.

Ericsson, Y. (1977) Cariostatic Mechanisms of fluorides: clinical observations. In: Cariostatic Mechanisms of Fluorides (W. E. Brown and K. G. König, eds.). Caries Res. 11 (Suppl. 1); 2-41.

Fanning, E. A., Leadbeater, M. M., and Somerville, C. M. (1975) South Australian kindergarten children: Fluoride tablet supplements and dental caries. Aust. Dent. J. 20; 7-9.

Forrester, D. J., and Schulz, E. M., Jr., eds. (1974) International Workshop on Fluorides and Dental Caries Reductions. School of Dentistry, University of Maryland, Baltimore, Maryland.

Forsman, B. (1974) Dental fluorosis and caries in high-fluoride districts in Sweden. Community Dent. Oral Epidemiol. 2; 132-148.

Forsman, B. and Ericsson, Y. (1974) Breastfeeding, formula feeding and dental health in low-fluoride districts in Sweden. Community Dent. Oral Epidemiol. 2; 1-6.

Garnier, P., Voegel, J. C., and Frank, R. M. (1976) Cinétiques de dissolution des cristaux d'hydroxyle-apatite synthétique et d'émail humain. Jour. Biol. Buccale 4; 323-330.

Grahnén, H., Myrberg, N., and Ollinen, P. (1975) Fluoride and dental age. Acta Odont. Scand. 33; 1-4.

Hollender, L., and Koch, G. (1969) Influence of topical application of fluoride on rate of progress of carious lesions in children. A long-term roentgenographic follow-up. Odontol. Revy 20; 37-41.

Horowitz, H. S., Heifetz, S. B., and Law, F. E. (1972) Effect of school water fluoridation on dental caries: final results in Elk Lake, Pa., after 12 years. J. Am. Dent. Assoc. 84; 832-838.

Hotz, P. (1969) Fluorkonzentrationen in der Mundflüssigkeit nach Verarbreichung von verschiedenen Fluorpräparaten. Med. Diss. Zürich.

Isaac, S. et al. (1958) The relation of fluoride in the drinking water to the fluoride in enamel. J. Dent. Res. 37; 218-325.

Jackson, D., and Weidmann, S. M. (1959) The relationship between age and the fluoride content of human dentine and enamel. A regional survey. Brit. Dent. J. 107; 303-306.

Koch, G., and Friberger, P. (1971) Fluoride content of outermost enamel layers in teeth exposed to topical fluoride application. Odontol. Revy 22; 351-362.

Kwant, G. W. et al. (1973) Artificial fluoridation of drinking water in the Netherlands. Netherlands Dent. J. 80; 6-27.

Lemke, C. H., Doherty, J. M., and Arra, M. C. (1970) Controlled fluoridation: the dental effects of discontinuation in Antigo, Wisconsin. J. Am. Dent. Assoc. 80; 782-786.

Marthaler, T. M. (1968) Die Kochsalzfluoridierung und Vergleich der kariesprophylaktischen Wirkung innerlicher Verabreichungsarten von Fluor. Dtsch. zahnärztl. Z. 23; 885-898.

Marthaler, T. M. (1969) Caries-inhibiting effect of fluoride tablets. Helv. Odont. Acta 13; 1-13.

Marthaler, T. M. (1971) Confidence limits of results of clinical caries tests with fluoride administration. Caries Res. 5; 343-372.

Marthaler, T. M. (1975) Improved oral health of school children of 16 communities after 8 years of prevention. II. Findings in different types of caries predilection sites. Helv. Odont. Acta 19; 2-12.

Marthaler, T. M. et al. (1978) Prevention of dental caries by salt fluoridation. Caries Res. 12, in press.

Mejia, R. et al. (1976) Estudio sobre la fluoruración de la sal. VIII. Resultados obtenidos de 1964 a 1972. Bol. Of. Sanit. Panam. 80; 67-80.

Medcalf, G. W. (1975) Six years of fluoridation on the gold-fields of Western Australia. Aust. Dent. J. 20; 170-173.

Moreno, E. C., Kresak, M., and Zahradnik, R. T. (1977) Physico-chemical aspects of fluoride-apatite system relevant to the study of dental caries. In: Cariostatic Mechanisms of Fluorides (W. E. Brown and K. G. König, eds). Caries Res. 11 (Suppl. 1); 142-171.

Mühlemann, H. R. (1965) Dietary fluoride and dental caries. Int. Dent. J. 15; 209-217.

Nicholson, C. R. and Mellberg, J. R. (1969) Effect of natural fluoride concentration of human teeth enamel on fluoride uptake in vitro. J. Dent. Res. 48; 302-306.

Rusoff, L. L. et al. (1962) Fluoride addition to milk and its effect on dental caries in school children. Am. J. Clin. Nutr. 11; 94-99.

Russell, A. L., and Elvove, E. (1951) Domestic water and dental caries. VII. A study of the fluoride-dental caries relationship in an adult population. Publ. Health Rep. 66; 1389-1401.

Russell, A. L., and Hamilton, P. M. (1961) Dental caries in permanent first molars after eight years of fluoridation. Arch. Oral Biol. (Spec. Suppl.) 6; 50-57.

Short, E. M. (1944) Domestic water and dental caries. VI. The relation of fluoride domestic waters to permanent tooth eruption. J. Dent. Res. 23; 247-255.

Silverstone, L. M. (1977) Remineralization phenomena. In: Cariostatic Mechanisms of Fluorides (W. E. Brown and K. G. König, eds.) Caries Res. 11 (Suppl. 1); 59-84.

Toth, K. (1976) A study of 8 years domestic salt fluoridation. Community Dent. Oral Epidemiol. 4; 106-110.

Toverud, G., Rubal, L., and Wiehl, D. (1961) The influence of war and postwar conditions on the teeth of Norwegian school children. Milbank Mem. Fund Q. 39; 489-539.

Van der Lugt, W., Knottnerus, D.I.M., and Young, R. A. (1970) NMR determination of fluorine position in mineral hydroxyapatite. Caries Res. 4; 89-95.

Weatherell, J. A. et al. (1977) Assimilation of fluoride by enamel throughout life of tooth. In: Cariostatic Mechanisms of fluorides (W. E. Brown and K. G. König, eds.) Caries Res. 11 (Suppl. 1); 85-115.

Yoon, S. H. et al. (1960) Distribution of fluorine in teeth and alveolar bone. J. Am. Dent. Assoc. 61; 565-570.

Ziegler, E. (1964) Bericht über den Winterthurer Grossversuch mit Fluorzugabe zur Haushaltmilch. Helv. Paed. Acta 19; 343-354.

3

Topical Fluorides in the Prevention and Arrest of Dental Caries

Erling Johansen and Thor O. Olsen

The concept of caries-prevention through topical fluoride treatments had its origin about 40 years ago (Volker and Bibby, 1941). Its validity was soon established by clinical testing (Knutson and Armstrong, 1943; Bibby, 1946) and has since been repeatedly verified. As a result of these efforts several topical fluoride formulations and application procedures have become part of accepted dental practice, and preparations designed for self-administration have won approval (Accepted Dental Therapeutics, 1973-1974). However, the caries-inhibitory effects derived from topical fluorides have seldom matched those obtained from consumption of fluoridated water.

The greatest benefit from topical fluorides has been shown to accrue to young persons who have newly erupted teeth and who reside in low-fluoride areas. In this population group which is more caries susceptible than adults the primary lesions originate in the enamel. With increasing age, however, the enamel undergoes posteruptive changes (Brudevold et al., 1960) which render the teeth less susceptible to caries. Older individuals experience gingival recession with root exposure, and root-surface caries becomes the primary affliction. While conventional topical fluoride regimens have little value in controlling this type of caries, daily topical fluoride treatments have been used with encouraging results in the prevention of cervical and root caries of patients suffering from radiation-induced xerostomia (Daly and Drane, 1972; Carl et al., 1972). However, the daily topical applications which are thought to be needed indefinitely make chronic fluoride intoxication a possible concern.

Figure 1. Frequency distribution of result of clinical studies on topical fluoride solutions.

NaF (2%)

1;4[28]
2;4[35]
1;2[11]   2;4[51]   1;4[11]   3;4[2]
3;4[7]   3;4[48]   1;6[11]   3;4[48]

SnF$_2$ 8%-10%

1;1[34]
1;1[53]   2;2[45]   1;1[47]
2;2[21]   3;1[51]   2;4[22]   2;1[30]
3;3[7]   3;6[15]   1;2[47]   2;4[6]   1;1[29]   2;4[30]   2;1[22]

AFP (1.23% F)

2;2[23] 1;1[16]   1;2[16]
2;2[23] 3;3[7]   2;2[17]   2;2[5]
2;4[1]   3;3[19]   3;3[19]   2;4[17]   2;1[53]   1;1[52]
3;3[7]   3;3[48]   3;4[48]   3;6[19]   2;4[6]   2;2[52]   1;2[24]

< 15%   25%   35%   45%   55%   65%   75% <

Reduction in DMFS Increments

Each study is represented by two digits separated by semi-colon. The first digit indicates the length of the study in years; and the second digit, the total number of topical applications. Superscripts refer to the list of references.

    Scrutiny of the literature on formulations of topical
fluoride agents, modes and length of time of administration
provides little clarification of what would constitute optimal treatment. A contributing difficulty in this regard is
the fact that the mechanism by which fluoride exerts its
cariostatic function is not established.

    There are two main categories of suggested mechanisms:
those which affect the environment of the tooth and, therefore, are transient; and those which affect the tooth itself
and may be persistent. If the transient effects are most
important, then the continued use of topical fluoride would
seem essential. On the other hand, if more permanent effects
are involved, then procedures which merely maintain the fluoride in the mineral might suffice after an adequate level of
protection has been established. In our opinion much of the
information in the literature suggests that the dominant
effect of fluoride is on the mineral, and the results of our
own studies on the mechanism of dental caries give further
support to this contention. We have, therefore, been in-
terested in developing and testing an alternative to the
programs which rely on continued daily applications of high
concentrations of fluoride.

Efficacy of Topical Fluorides

    The positive results obtained in the early clinical
trials of professionally applied NaF solutions (Knutson and
Armstrong, 1943; Bibby, 1946; Galagan and Knutson, 1947) stimulated much interest in this approach to caries prevention.
There has been continuing search for improved formulations
and for more efficient and economical methods of administration of the fluoride. The results of these efforts include
the development of stannous fluoride (SnF_2) preparations
(Muhler et al., 1954; Muhler, 1958), acidulated phosphate
fluoride (APF) solutions and gels (Brudevold et al., 1957;
1963) and monofluorophosphate (MFP) dentifrices (Naylor and
Emslie, 1967; Mergle, 1968a, 1968b; Møller et al., 1968).
Since authoritative reviews have recently been published
elsewhere (Brown and König, 1977; Forrester and Schulz, 1974),
only an overview of these studies is included here.   '

    The data presented in Figure 1 are representative of
findings in studies on topical fluoride solutions. The results are widely scattered, but the median caries reduction
is in the 30-40% range for each of the three preparations
considered, which is less than the benefits commonly observed
with water fluoridation. Only a few studies have tested the
effectiveness of these preparations in communities with fluoridated water supplies. The results (not included in Figure

Figure 2. Frequency distribution of reduction in DMFS increments in clinical trials with fluoride dentifrices and mouthrinses.

	< 15%	25%	35%	45%	55%	65% <
SnF$_2$				1/2;u[37]		
				1/2;u[40]		
				3/4;u[54]		
				1;u[12]		
	1;u[20]	1;u[36]	1/2;u[36]	1;u[40]		
	1;u[26]	1½;u[36]	3/4;u[54]	1;u[42]		
	2;u[20]	2;u[36]	1;u[12]	1½;u[54]		
	2;u[26]	3;u[20]	1;u[37]	1½;u[54]		
	2;u[31]	3;u[36]	2;u[13]	2½;u[55]	1/2;u[42]	
	3;u[43]	2;s[25]	2;u[38]	2½;u[55]	1;u[39]	1/2;s[44]
	1;s[20]	2;s[25]	2;s[49]	2/3;s[3]	2/3;s[3]	2/3;s[4]
	2;s[20]	3;s[20]	2;s[49]	2;s[13]	1½;s[4]	3/4;s[44] 1/2;u[41]
PO$_3$F		2;u[10]				
		3;u[32]				
		2;s[31]				
	3;u[43]	2½;s[33]	2;s[50]			
NaF Mouthrinse		2;w[18]		2;f[18]		
		2;f[51]		2;d[51]		
	5;f[27]	3;f[27]		3;d[46]		

Reduction in DMFS Increments

Each dentifrice study is represented by one digit or a fractional number, reflecting its length in years, and a letter; u for unsupervised or s for supervised. Each mouthwash study is represented by a digit, indicating length of study in years, followed by a letter, indicating frequency of use as follows; d-daily; w-weekly and f-fortnightly. Superscripts refer to the list of references.

1) for sodium fluoride (Downs and Pelton, 1950; Galagan and Vermillion, 1955) and stannous fluoride (Horowitz and Heifetz, 1969) have generally been discouraging in that little or no benefit was found.

Fluoride-containing varnishes have been tested in a few clinical studies. The results have been inconsistent, ranging from better than 70% reduction in DMFS increment (Koch and Petersson, 1975) to no effect (Murray et al., 1977). Both extremes in the range of results were obtained with the same preparation (Duraphat).

Many other fluoride compounds have shown promise when tested in vitro or in animals, but have been subjected to little or no clinical testing. Agents which have shown promise in preliminary clinical testing include TiF_4 (Reed and Bibby, 1976), and FeF_3 (Berggren, 1967).

Prophylactic pastes have not given convincing results when used as the only source of topical fluoride, but good results have been obtained when prophylaxis with fluoride-containing paste was followed by topical application of fluoride solution, with the best results (58% reduction in DMFS increment) when the patients also used a SnF_2 dentifrice daily (Davies, 1974).

Vehicles for providing fluoride posteruptively to the teeth by self-application include dentifrices, mouthrinses and lozenges. As illustrated in Figure 2 for SnF_2 and MFP dentifrices the results of clinical tests of these agents vary widely. The results for SnF_2-containing dentifrices range from no (or a slightly detrimental) effect to well over 60% reduction in DMFS increment, with a median in the 25-30% range. For MFP dentifrices the studies are far fewer, but the results are also closer together, with a median of about 20%. A few investigators have tested, with good results, the effectiveness of amine fluoride (35% reduction in DMFS; Marthaler, 1968) and NaF-containing dentifrices (40-50% reduction, Koch, 1967).

Mouthwashes containing sodium fluoride have provided only minimal benefit when used less frequently than once a week. A 45% reduction in DMFS increment was obtained in two studies, one using 0.2% NaF solution once a week, and the other, 0.05% solution daily (Koch, 1969). The magnitude of the effect may depend on the age of the participants; a procedure which gave 43% reduction (DMFS) in ten-year-old children after two years gave only a nonsignificant 16% reduction in six year olds (Horowitz et al., 1971).

Brushing with amine fluoride solution supplemented with fluoride mouthrinse has given good results ($\simeq$ 50% reduction in DMFS) in studies in Switzerland (Marthaler et al., 1970). Other studies with brushing solutions have given reductions in DMFS of 25% with NaF (Berggren and Welander, 1960) and 10-15% with APF (Bullen et al., 1966; Horowitz and Heifetz, 1968).

The great variability in results found for most topical procedures is not readily explained from the information provided in the reports. Variables such as patient selection, details of the techniques used and patient age distribution are likely to have played a role, but information regarding their exact contribution is not available. Additional, often unidentified variables, such as prior fluoride exposure and dietary habits, may also have affected the results. Overall, it is, therefore, difficult to generalize from the reported data concerning effectiveness to be expected of these preparations and treatment modes.

Self-applications of topical gels at frequent intervals were first tried by Englander et al. (1967). In that study school children in a nonfluoridated area used custom-fitted plastic trays to apply a 1.1% NaF gel (neutral or acidulated) every school day for up to two years. The procedure resulted in a 75-80% reduction in new decayed and filled surfaces. Two years after the daily applications were discontinued, the fluoride groups still showed lower DMFS increments (by 70%) than did the control groups (Englander et al., 1969). In an extension of the study to a fluoridated area, however, the same investigators found only a 29% reduction in DMFS increment after two years of gel use and concluded that the procedure would be impractical in such areas (Englander et al., 1971).

The technique of daily self-application of fluoride gel in custom-fitted plastic trays has also been used in a number of studies on prevention of radiation caries. The original work by Daly and Drane (1972), and a subsequent study by Dreizen et al. (1976) using the same techniques have shown encouraging results. Two alternative approaches for frequent fluoride applications have also been used with irradiated patients: Shannon's group developed a stannous fluoride-glycerol preparation which the patients apply by brushing and allow to dry on the teeth every evening (Wescott et al., 1975), and Regezi et al. (1976) reported on a program where the patients rinsed with a 3% NaF solution followed by one liter of a $NaHCO_3$ - NaCl solution three times a day.

Daly and Drane (1972) and Daly et al. (1972) reported that of 67 irradiated patients using a 1% NaF gel on a daily basis for up to 6-1/2 years only 30.7% developed caries. The corresponding percentage for the control group was 68. Dreizen et al. (1976), using the technique of Daly and Drane, found DMFT increments of 0.07 per month in the treatment group as compared to 1.24 in the control group using a placebo gel.

The above results obtained with daily use of fluoride gels or solutions have been encouraging, but this frequency of use of preparations with fluoride ion concentrations at or above 0.5% can hardly be recommended for the normal population. The groups using these techniques have reported no data on fluoride absorption, but since 5 ml of the preparation contains at least 25 mg fluoride, the possibility of ingesting harmful quantities of fluoride does exist (see Taves; Jowsey et al., this monograph).

Mechanisms of Action of Topical Fluorides

Much effort has been devoted to determination of the mechanism or mechanisms of the cariostatic action of fluorides (for detailed information, see Brown and König, 1977). Unfortunately, there is evidence for several mechanisms without a clear indication of their relative importance. The possible mechanisms of action fall into two categories, those in which fluoride affects the environment of the tooth and those in which the fluoride affects the properties of the tooth itself.

Interactions between fluoride and the oral environment have been described for concentrations down to the part-per-million range. These include effects on bacterial growth (Marquis et al., 1976), glycolysis (Bibby and Van Kistern, 1940), glycogen synthesis (Weiss et al., 1965), acid production (Wright and Jenkins, 1954), production of extracellular polysaccharides thought necessary for plaque adhesion to tooth surfaces, and solubility of calcium phosphate deposits within the plaque (Kleinberg, 1970; Broukal and Zajicek, 1974). All these effects are likely to occur after a topical fluoride treatment, at least when the fluoride concentration of the preparation is high. However, as plaque is removed by abrasive forces and oral hygiene procedures, the fluoride incorporated in these deposits during treatment will disappear. Furthermore, the amount of fluoride deposited in the teeth by a topical application is limited, so the tooth cannot supply fluoride to the plaque for extended periods of time. Effects of topical fluoride on plaque are, therefore, likely to be of short duration (in the order of less than a day to a few weeks, depending on the mechanism considered), and effective prevention of caries through these mechanisms would require frequent -- perhaps daily -- applications.

Several mechanisms through which fluoride may have a cariostatic effect directly on the teeth (enamel) have been suggested. These include enhancement of the rate of salivary remineralization of beginning lesions, reduction of adsorptive forces necessary for bacterial deposits to adhere to the teeth and reduction of the solubility of the tooth mineral. The effects in this category should have the advantage of greater permanence than the plaque effects.

Pigman et al. (1964) and Koulourides et al. (1965) showed that fluoride concentrations as low as 1 ppm enhanced the rate of remineralization (rehardening) of softened enamel in vitro by saliva and synthetic mineralizing solutions. Thus, the presence of relatively low concentrations of fluoride from external sources or from dissolving mineral in an early lesion, may enhance the physiological repair sufficiently to prevent a clinical lesion from developing. Also, other evidence (Brown et al., 1977) indicates that fluoride may promote conversion of dicalcium phosphate to (fluoride-containing) apatite. Since hydroxyapatite surfaces exposed to a solution of pH less than approximately 4.5 appear to undergo phase transformation to dicalcium phosphate (Gray et al., 1962); the enhancement of its convertion to apatite would help reduce the loss of mineral during each acid-base cycle of the carious process.

Hydroxyapatite is known for its ability to adsorb a great variety of compounds to its surfaces (it is a commonly used adsorbent in chromatography). Fluoroapatite and fluoride-containing hydroxyapatite reportedly show weaker adsorptive properties (Ericson and Ericsson, 1967; Tinanoff et al., 1976), and fluoride-treated enamel, less wetting (Glantz, 1969) by polar liquids than untreated enamel. Therefore, if the external mineral surfaces of the tooth were converted to fluoride-containing apatite, the adhesion of bacterial plaques might be diminished, facilitating their removal and hindering their growth.

The solubility-reduction theory (Volker, 1939) has been the subject of considerable controversy regarding its importance at the fluoride concentrations found in the dental tissues (see Brown et al., 1977). It is, however, supported by many studies (Isaac et al., 1958; Brudevold and McCann, 1968; Cutress, 1972) which have shown that small increases in the amount of fluoride in enamel markedly reduced the (equilibrium) solubility.

Further support for the hypothesis that fluoride stabilizes the mineral phase is obtained from coordinated ultrastructural and chemical studies on sound and carious dental tissues. These studies show that, at three different levels, surface mineral exhibits greater resistance to caries, or acid dissolution, than does bulk mineral. The surface of enamel to a depth in the order of 30 μm (Darling, 1963) and cementum (Furseth and Johansen, 1968) remain relatively unaffected by the caries process while underlying bulk tissue is extensively demineralized. At the cellular level, the surfaces of enamel rods (Johansen, 1963; Glas et al., 1965) and the pericanalicular zone of dentin (Johansen and Parks, 1961) frequently remain less severely affected by the caries process than does the adjacent bulk tissues, i.e., rod core or intercanalicular areas. Finally, at the crystallite level, electron micrographs of carious enamel show that shell-like remnants of crystal surfaces remain in extensively demineralized areas (Johansen, 1963; 1965). In the absence of a difference in composition the small fragments would be expected to be more soluble than the much larger crystallites of the underlying normal tissue (Neuman and Neuman, 1958). In the enamel lesion, however, acid diffuses past the fragments to dissolve underlying mineral. This observation indicates that the remaining fragments have a composition which lends stability in the carious environment. Also, in carious dentin and cementum some plate-like and tablet-shaped crystallites remain apparently unaffected by the acids while crystallites of adjacent sound tissues are being dissolved (Johansen, 1963; Furseth and Johansen, 1970). These observed differences in solubility appear to coincide with differences in fluoride content of the mineral.

The surfaces of all three dental tissues (enamel, dentin and cementum) have been shown to acquire fluoride, or high-fluoride mineral, upon exposure to oral fluid (enamel, Brudevold et al., 1956; dentin and cementum, Olsen and Johansen, 1970, 1973; see Table 1). This phenomenon reflects the nature of the normal oral fluid-apatite system. The mechanism is also reflected at the cellular level where the surfaces of the structural units have been postulated to contain more fluoride than the bulk because they are more assessible to fluoride diffusing into the enamel from the surface (Johansen, 1965). In dentin the source of fluoride is the pulpal fluid (Johansen and Parks, 1961). Finally, mineral persisting in extensively demineralized regions of the carious lesions is known to contain more fluoride than the sound tissue (enamel, Johansen, 1963; dentin, Johansen and

Table 1.   Fluoride Content of Ashed Dental Tissues.   Comparison of Carious and Exposed with Sound, Unexposed Tissue

Tissue	Mean*	SD	N
Exposed vs. Nonexposed			
Dentin, Exposed	230	96	
Nonexposed	130	67	11
Cementum, Exposed	2200	510	
Nonexposed	1600	440	10
Carious vs. Sound			
Enamel, Carious, bulk	500	490	
Sound, bulk	63	46	25
Dentin, Carious, acute	1800	1200	
Sound	290	230	24
Cementum Carious	3800	1100	
	2500	760	10

*ppm of ash weight.   Olsen and Johansen (1970; 1972; 1973).

Nordback, 1962; Levine, 1973; cementum, Olsen and Johansen, 1972).

In addition to the qualitative aspects discussed above, the levels of fluoride found in mineral of extensively demineralized carious lesions should be a useful guide to the levels which would render the normal tissues resistant to caries attack. These levels (Table 1) were found to be about 500 ppm for subsurface enamel and 2000-3000 ppm for dentin and cementum. The difference between enamel and dentin (or cementum) is related to the difference in the surface to volume ratio of the crystallites. For enamel the data are confirmed by our finding that moderately fluorosed enamel, which is highly caries resistant in spite of its structural defects (porosities), contained 3500 ppm F in the surface layer (at 5 μm) and 500 ppm at 100 μm (Table 2).

It is possible that there is more than one important mechanism of action of fluoride and that they are not utilized to the same degree with the various topical fluoride procedures. Daly and Drane (1972) reported that dental conditions deteriorate rapidly upon discontinuance of daily use of the gel in xerostomia patients. This observation may indicate that bacterial or plaque-related effects were important in that study. However, because these patients have limited saliva, other possibilities do exist. The fluoride may have been lost or may be ineffective without the

Table 2.   Enamel Fluoride in Moderately Fluorosed and Normal
           Teeth*

	depth (μm)[†]					
	5	10	25	50	100	No.
Moderately Fluorosed (Colorado)	3470** ±2070	2550 ±1380	1330 ±410	780 ±160	490 ±67	29
Normal-1 ppm F (Rochester, N.Y.)	1210 ±470	660 ±300	280 ±120	170 ±65	– –	39

*Normal teeth from 1 ppm F area
**Mean and SD
[†]F concentration at indicated depths obtained by interpola-
 tion on smooth curves fitted to data points.   Enamel
 sampled by successive acid-glycerol etchings   (Olsen and
 Johansen, 1978).

remineralization provided by saliva.   The latter two explanations seem plausible in light of our studies mentioned above
and offer the possibility of a new approach to caries prevention.

A New Approach in the Use of Topical Fluorides

The maintenance of the tooth in the oral environment
involves a dynamic multifaceted system including dissolution,
remineralization and ionic exchange.   Therefore, in order to
eliminate the need for daily applications of large amounts
of fluoride for extended periods of time, we have designed
a preventive program which is a combination of several procedures.   Most of these procedures have been used alone
without spectacular results.   Together they hold the promise, however, of more than the partial control or the unacceptable risks which have been available.

The measures and their rationale are as follows: short-
term intensive use of topical fluoride gel to raise the
fluoride concentrations of both the surface and underlying
tissues to levels which would be expected to give protection
against caries (i.e., at least to the levels of fluoride
found in carious tissue and, in the case of enamel, in
moderately fluorosed teeth); daily use of fluoride toothpaste to maintain the fluoride levels obtained; stimulation
of salivary secretion, by use of a special chewing gum to
enhance the protective mechanisms inherent in the saliva;
intermittent use of supersaturated solution of calcium,
phosphate and fluoride to augment the physiological

maintenance system and provide new nucleation sites for
mineral growth; brushing and flossing, supplemented by a
new dental hygiene procedure of cleaning with cotton swabs,
prior to the use of the fluoride gel or the mineralizing
solution to allow the chemicals direct contact with the
tooth surfaces, and also to decrease the amount of acid produced; and finally, rinsing with water after ingestion of
carbohydrates to further reduce the acid attack rate.

Preliminary Studies

The relative effectiveness of a 2.2% NaF solution and
several gel formulations (APF (1.23% F), 2.2% NaF, 1.1% NaF,
and 1% NaF (M.D. Anderson)was first assessed in vitro. The
best results in terms of fluoride deposition and depth of
penetration were obtained with the APF gel followed by the
most concentrated of the NaF gels. These two preparations
were, therefore, selected for further testing which included
comparison of 15-minute, 30-minute and 60-minute single
treatments. The results showed that extension of the treatment time from 15 to 60 minutes with a single application
had little effect on uptake and penetration with either formulation. Presumably, reaction products, such as CaF_2 which
forms quite readily under the conditions of these treatments,
result in both inhibition of further fluoride uptake in the
superficial layer and blockage of fluoride diffusion into
the interior of the enamel.

In order to circumvent this blockage, repeated applications were tried separated by exposure to a supersaturated
calcium phosphate solution ("artificial saliva") which would
cause dissolution of CaF_2 and facilitate formation of fluoride-containing apatite. The results obtained in vitro
showed that a sequence of 12 5-minute treatments separated
by two-hour exposures to fluoride-free artificial saliva
was superior to the 60-minute single treatment (Table 3)
and suggested that frequently repeated topical applications
might be an effective treatment mode (Olsen and Johansen,
1977).

To establish the required treatment time, extracted
teeth were inserted in a partial denture and treated intra-
orally with APF (1.23% F) gel once a day. Teeth were removed for analysis at one-week intervals from one to six
weeks after the initiation of the treatment. Analyses of
these teeth indicated that treatment times of three weeks or
more would result in fluoride levels in the enamel adequate
for caries protection. Results of a similar test --
carried out for four weeks only -- with the NaF (1% F) gel
indicated that this treatment time would give results

Table 3. Fluoride Content (ppm) of Enamel after Treatment
          In Vitro with Fluoride Gel

Treatment	depth (µm)[†]				No.
	5	10	25	50	
APF[+] (1.23% F)					
1x60 min	4100	1300	660	260	4
	±1200	±350	±120	±30	
12x5 min	8600	4600	1230	370	4
	±2500	±1400	±330	±170	
NaF					
1x60 min	3900	1900	670	330	4
	±1500	±1100	±310	±110	
12x5 min	8500	4900	1900	810	4
	±1100	±1600	±460	±220	

*Mean ± SD
[+]APF: Acidulated phosphate fluoride gel
 NaF: Sodium fluoride gel
[†]For details see Table 2 (Olsen and Johansen, 1977).

approaching those desired (compare Tables 4 and 2). (In the
final design, fluoride gel was used twice a day for two weeks
and then daily for two weeks.

The remineralizing solution could, for practical reasons,
be used by the patients only for limited periods of time.
Therefore, a degree of supersaturation was chosen which
approached that causing immediate precipitation upon preparation. This should help accelerate the nucleation process
and maximize the amount of mineral deposited from the solution per unit time. The solution contained 5 mM Ca, 3 mM
PO_4 and 0.25 mM (5 ppm) fluoride. It was stabilized by
NaCl, as suggested for such systems by Koulourides et al.
(1968), at pH 7.0. Studies of the solution were carried out
in vitro to provide information on its ability to react with
existing tooth mineral as well as to form new mineral within
or on sound and decayed dental tissues. The uptake of Ca,
PO_4 and F was determined for carious dentin pieces and for
powdered enamel and dentin, both normal and partly demineralized.

Typical uptake curves for F and Ca from the mineralizing solution are shown in Figure 3 and 4, respectively.
The percentage of PO_4 removed from the solution approximated
that of calcium. Control solutions (containing no dental
tissue) showed no appreciable change in F, Ca and PO_4 concentration over the 90 minutes illustrated in the figures
(T. Olsen and E. Johansen, unpublished).

Ultrastructural studies showed that triturated speci-

Figure 3. In *vitro* uptake of fluoride from remineralizing mouthwash by dental tissues. The curves show percentage of original fluoride concentrations remaining in solution as a function of time.

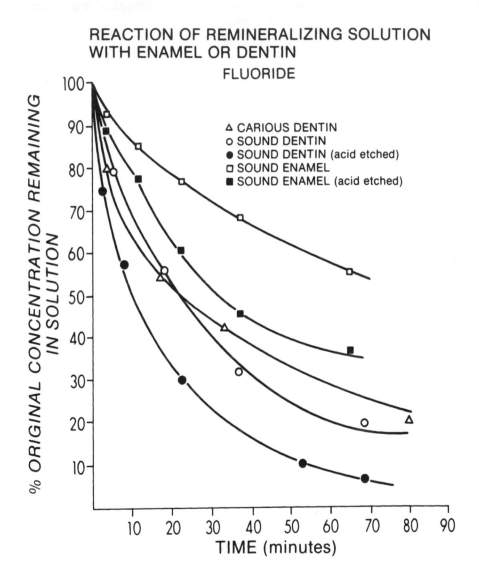

REACTION OF REMINERALIZING SOLUTION
WITH ENAMEL OR DENTIN
FLUORIDE

△ CARIOUS DENTIN
○ SOUND DENTIN
● SOUND DENTIN (acid etched)
□ SOUND ENAMEL
■ SOUND ENAMEL (acid etched)

Figure 4. In vitro uptake of calcium from remineralizing mouthwash by dental tissues. The curves show percentage of original calcium concentrations remaining in solution as a function of time.

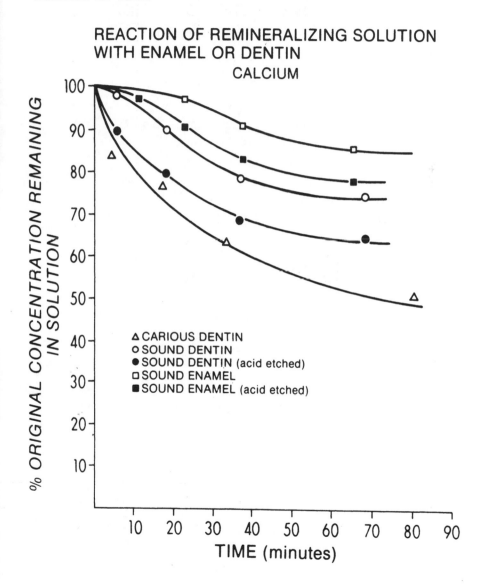

REACTION OF REMINERALIZING SOLUTION WITH ENAMEL OR DENTIN

CALCIUM

Table 4.   Fluoride Content (ppm) of Enamel of Nonvital Teeth
(Denture) after Treatment In Vivo with Fluoride
Gel

Treatment	depth* (μm)					No.
	5	10	25	50	100	
APF (1.23% F)						
1 week	1500	800	380	180	110	1
2 weeks	8200	7000	3400	980	-	1
3 weeks	6300	5600	2200	700	-	1
4 weeks	13500	9800	4600	2200	-	1
5 weeks	16000	10000	6000	3500	-	1
6 weeks	11000	8000	3000	2300	1400	1
NaF (1.0% F)						
4 weeks	6300 ±1300**	3400 ±900	1800 ±550	1100 ±420	350 ±130	7

*For details see Table 2
**SD.   (T.Olsen and E. Johansen, unpublished)

mens of enamel and dentin treated with the remineralizing
solution for two hours contained crystallite species not
native to the tissues, but similar in morphology to deposits
formed spontaneously in the solution on standing for several
hours (Figures 5-9; T. Olsen and E. Johansen, unpublished).
Teeth with carious lesions were also treated in vivo while
mounted in a denture.   The patient brushed his teeth only
with water and used the remineralizing solution as a mouthwash twice a day for the four weeks of the experiment.   The
treated carious dentin was embedded and sections were
studied in the electron microscope (Figure 12) and compared
with sections of carious dentin treated in vitro (Figure 11)
and precipitates from the solution (Figure 10; E. Johansen,
R. Hill and T. Olsen, unpublished).   Sections of both types
of treated carious dentin showed a high degree of mineralization with clusters of crystallites extending into the lumina
of the canals.   The morphology of these crystallites was
similar to that of crystallites observed in spontaneous
precipitates from the solution, but has not been observed in
untreated specimens of carious dentin.   These findings
suggested that the mouthwash was able to induce mineralization in dental tissues both in vitro and in vivo.

Clinical Study

Clinical testing of the preventive program commenced
in 1973 when Dr. Sidney Sobel* enlisted our cooperation in

*Radiation Oncologist, Highland Hospital, Rochester, N.Y.

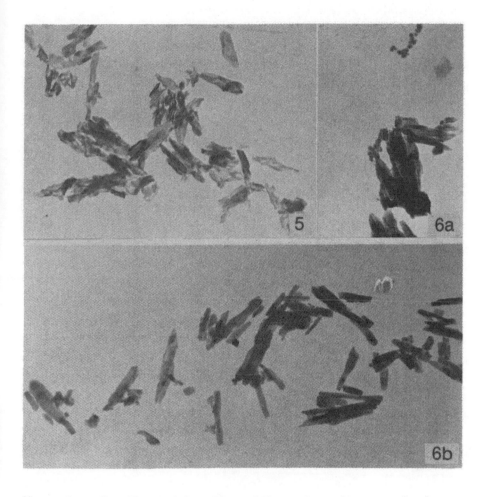

Human Enamel. Treated <u>in</u> <u>vitro</u> with remineralizing solution.
<u>Figure 5</u>. Triturated sample of tissue after partial demineral-
ization with 2 M perchloric acid saturated with dicalcium phosphate (X 25,000). <u>Figure 6a,b</u>. Triturated sample of tissue
after demineralization and subsequent exposure to remineralizing
solution. Both newly formed globular precipitates and dagger-
shaped crystallites are found adjacent to tissue (X 20,000) (a).
Individual crystallites appear more electron dense and their
shapes are more regular than those not treated with remineralizing solution (ref. Figure 5) (b) (X 25,000).

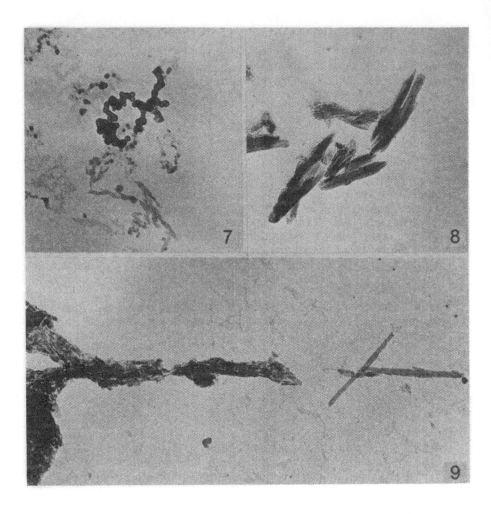

Human Dentin. Treated in vitro with remineralizing solution.
Figure 7. Precipitate from remineralizing solution showing ag-
gregates of globular bodies and beginnings of elongated crystallites (X 40,000). Figure 8. Dagger-shaped crystallites found
on carious dentin treated with remineralizing solution
(X 27,000). Figure 9. Long rod-shaped crystallites found adjacent to sound dentin after exposure to remineralizing solution (X 35,000).

the oral management of oncology patients experiencing radiation-induced xerostomia (Johansen et al., 1975; J.Am.Med. Assoc., Vol.234, pp. 577-8, 1975). Subsequently, the study was expanded to include nonirradiated patients with high caries susceptibility. This report will present results for several groups of patients.

Materials and Methods

All patients in the study were referred by their physician or dentist. This report deals with 155 patients all of whom had been followed for at least one year. Of these, 118 had been irradiated, and the others had a variety of problems, such as; xerostomia due to chemotherapy, Sjøgren's syndrome, scleroderma and other causes; defective tooth enamel due to hypoplasia and amelogenesis imperfecta; or simply a history of recurring caries. Patient age ranged from 7 to 80 years.

The oral health status of each patient was usually assessed by three dentists, including the patient's private dentist, at the start and subsequently at approximately six-month intervals. In most cases, carious lesions were restored prior to initiation of the preventive program, but, eighteen patients with (128) active carious lesions were started on the preventive program without restorative work. The effect of the preventive procedures on these lesions was judged by the color and hardness of the carious tissues at intervals throughout the study. The principal criterion used for diagnosis of a carious lesion was that it appeared progressive. Some lesions were filled by the private dentist without consultation and may not have been progressive. These were nevertheless included in the cumulative results.

The preventive procedures were explained and demonstrated to each patient either individually or in groups, and written instructions were also provided. Briefly stated the procedures were as follows:

Oral hygiene: Flossing, standardized tooth brushing with fluoride toothpaste, and cleaning of tooth surfaces with cotton swabs.

Topical fluoride applications: Self-administration of fluoride gel by means of custom-made trays (Englander et al., 1967) of soft plastic (Mouthguard[R] material). Patients whose dentition was in good condition were prescribed (APF) gel (1.23% F); patients with active carious lesions, extensive old restorative work or sensitivity to the acid formulation were prescribed neutral

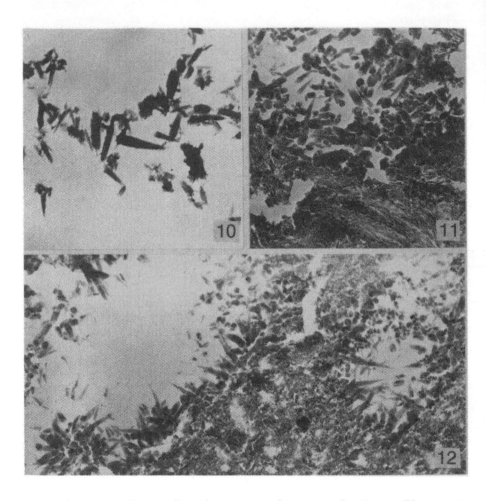

Human Dentin. Treated with remineralizing solution. Figure 10.
Dagger-shaped crystallites found in precipitate from remineralizing solution without exposure to tissue (X 30,000). Figure
11. Section of carious dentin treated in vitro with remineralizing solution. Dagger-shaped crystallites similar to those
seen in Figure 10 are evident (X 20,000). Figure 12. Section
of carious dentin from tooth worn in partial denture for 4
weeks and treated with remineralizing solution twice a day as
prescribed. The carious tissue is remineralized with crystallites resembling those found in the precipitate from the remineralizing solution. Large numbers of dagger-shaped crystallites
protrude into the lumen of the dentinal canals from the border-
ing tissue (X 27,000).

sodium fluoride gel (1% F, prepared in our laboratory). The treatment schedule consisted of two five-minute applications per day for two weeks followed by daily applications for an additional two weeks. Following each treatment, the mouth was to be thoroughly rinsed with water to remove residual gel and prevent swallowing of fluoride. In children under the age of ten, the initial treatment schedule was shortened, but treatment was resumed after eruption of new teeth. For a few adult patients who did not follow the oral hygiene procedures conscientiously, a limited number of booster treatments were prescribed on an individual basis. For the 18 patients with untreated lesions, the length of the initial treatment schedule was 4-8 weeks, and some continued to use the gel on a schedule adjusted to the individual's need.

Remineralizing mouthwash: A two-minute rinse was prescribed after each topical fluoride application. When the fluoride gel treatment was completed, the patients were asked to continue using the solutions twice a day after toothbrushing. Two stock solutions were stored separately and mixed in proper volumes immediately before use.

Salivary gland stimulation: A non-sweetened gum was formulated and prescribed for patients with xerostomia to stimulate salivary secretion.

To assess the amount of fluoride ingested during treatment and the effects of the treatment procedures on the composition and structure of the teeth the following laboratory studies were performed:

Urinalysis for fluoride: Two 24-hour urine samples were collected from about 50 patients for each of the following periods: before topical treatment, during the period of two treatments per day, and during the period of daily treatments. Fluoride analyses were carried out with an Orion fluoride electrode after adjustment of pH and ionic strength. Creatinine was also determined (by colorimetry) and used as an indication of the completeness of the urine collection (Cloutier and Johansen, 1977). Results from collections which appeared incomplete were not included in the tabulations.

Chemical analysis of teeth: Treated teeth were obtained when deciduous teeth were shed, and when teeth fractured or were extracted. Intact areas on the enamel were sampled repeatedly with a window technique and an acid-glycerol mixture similar to that described by Brudevold et al. (1975). Carious and exposed dentin and cementum were sampled by grinding, and all samples were analyzed for

Table 5. Increments of caries and replaced restorations in control, partial compliance and prevention groups.

Patient Group	1st year						2nd year						3rd year					
	Patients			Teeth			Patients			Teeth			Patients			Teeth		
	#	%C*	%R	#	%C	%R	#	%C	%R	#	%C	%R	#	%C	%R	#	%C	%R
Control†	26	92.3	-	403	72.2	-												
Partial Compliance†	10	90.0	50.0	244	20.5	7.8												
Compliance Irradiated	108	10.1	24.8	2580	0.7	2.3	58	13.8	34.5	1452	1.4	2.8	31	25.8	25.8	802	1.2	1.4
Other Cancers	7	14.3	14.3	167	0.6	3.0	3	0	0	87	0	0						
Salivary Problems	12	16.7	8.3	286	0.7	2.5	2	0	0	41	0	0						
Defective Tooth Struct.†	4	0	25	107	0	0.9	2	0	0	57	0	0						
Highly Caries† Suscept.	14	14.3	28.6	359	0.8	2.8	1	0	0	28	0	0						
All Compliance Groups Comb.	145	11.0	23.3	3499	0.7	2.3	66	12.1	30.3	1665	1.2	2.5	31	25.8	25.8	802	1.2	1.4
All Prevention Groups**	155	16.1	25.2	3743	2.0	2.7												

*%C refers to percentage of patients (or teeth) with carious lesions. %R is the percentage of patients (or teeth) requiring replacement of defective restorations.

†Mean DMFT for the highly caries susceptible group was 21 at start.

†All control and partial compliance patients received head and neck irradiation.

**Includes partial compliance group.

fluoride using the Orion electrode after buffering and adjusting the ionic strength of the sample solutions. Calcium and phosphate were determined colorimetrically, and the results were used to estimate the ash weight of the samples.

Electron microscopy of precipitates and dental tissues: Precipitates from the remineralizing mouthwash were transferred to carbon-coated grids and rapidly washed with deionized water before study. Samples of dental tissues from the patients were obtained by removal of specimens from teeth retained by the patient or from exfoliated, fractured or amputated teeth. Some specimens were gently triturated in distilled, deionized water and deposited on carbon-coated grids for study. Other tissue samples were fixed in glutaraldehyde and embedded in Spurr's medium. Sections were prepared with a diamond knife and studied without staining for characterization of the crystallite species of the mineral component.

Results

The caries experience of patients in the various groups is presented in Table 5. It can be seen that over 90% of the irradiated control patients (72% of the teeth) had developed lesions within a year. In the partial compliance group (those who admitted having complied only marginally with the procedures of the initial stages of the program, including the use of topical fluoride gel), 90% of the patients developed carious lesions involving 20% of the teeth, but only a few patients in this group developed cervical lesions typical of salivary deficiency.

These results differ markedly from those of the compliance group of irradiated patients. Ten percent of these patients developed lesions which needed filling or were filled by their private dentist, and only two patients developed cervical lesions. About one percent of the teeth in this group became carious during each year on the program, and these lesions occurred mainly in the pits and fissures of younger patients. In contrast, in the partial compliance group many lesions were interproximal.

The decay process was arrested over a period of from four to eight weeks in all but three of the 18 patients started with active lesions; many lesions continued to show increasing hardness to the touch of a sharp explorer in subsequent weeks. Some rehardened lesions were later filled by the patient's dentist, for esthetic reasons or because the patients were unable to keep them clean. Other lesions were

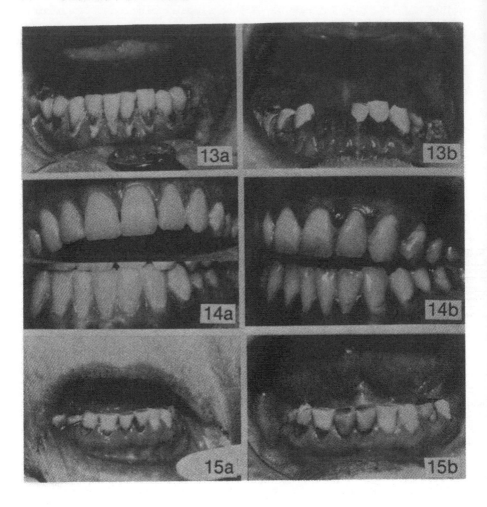

Dentition of patients irradiated to head and neck. Figure 13a.
Mandibular dentition of 52-year-old male patient before radiation therapy in treatment of nasal carcinoma. Figure 13b. Same
patient 11 months later; without preventive program all teeth
were extensively decayed. Figure 14a. Dentition of 19-year-old
female patient 5 months after radiation therapy in treatment of
Hodgkin's disease. Most teeth demonstrate white spot carious
lesions. Figure 14b. Same patient after 2 years on preventive
program; all lesions have remained arrested. Figure 15a. Mandibular teeth of 63-year-old female patient 21 months after radiation therapy for lymphoma. Several teeth show extensive cervical caries. Figure 15b. Same dentition 2 years later. All
lesions were rehardened by preventive procedures; some fillings
were subsequently placed for cosmetic reasons.

left without restoration for periods up to four years. No periapical abscesses developed in the teeth treated by this procedure.

Table 5 further shows that the results for the other groups on the preventive program compare favorably with those of the irradiated compliance group. Overall about one percent of the (3499) teeth of 145 patients developed lesions in one year. The data for the succeeding two years show this rate to be rather constant. Inclusion of the partial compliance group into the overall experimental group changes the values for the first year slightly as follows: two percent of the teeth developed caries, and 2.7% of the teeth needed replacement of restorations.

Visual evidence of the destructive nature of the carious process in the various groups and of the beneficial effects of the preventive program is presented in Figures 13-20. Comparison of Figures 13a and 13b shows the destruction of an irradiated patient's teeth over a period of 11 months in the absence of preventive procedures. The photograph does not, however, convey the fact that most of the teeth (Figure 13b) had become as soft as rubber. This is a typical consequence of a high dose of radiation in an older individual with exposed dental roots. The lesions depicted in Figure 14a are often observed in the cervical enamel of younger patients, in this instance five months postirradiation. The enamel lesions are, however, easily arrested by the preventive program, as illustrated in Figure 14b taken about two years later. These lesions need not be restored. The advanced cervical lesions, involving both enamel and dentin, depicted in Figure 15a developed in less than two years without prevention in a patient with moderate radiation exposure. All these lesions were rehardened, but some were subsequently restored for esthetic reasons (Figure 15b).

The benefit derived from the chemical treatment program is illustrated in the dentition depicted in Figures 16a,b. When this 16-year-old male patient was first seen and started on the preventive system, shortly after conclusion of radiation therapy, extensive decalcification and softening of the labial enamel had already taken place. After a 3-1/2 year postirradiation period, these teeth were still intact.

Those patients complying with all aspects of the preventive system showed excellent oral conditions and healthy teeth throughout the period of observation. Figure 17 illustrates the healthy dentition of a 52-year-old man with exposed roots who during approximately four years on the

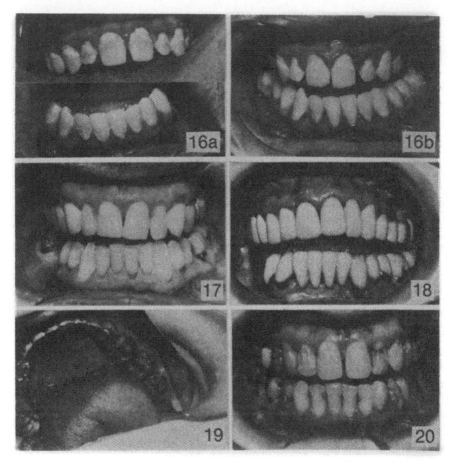

Dentition of irradiated and nonirradiated patients.   Irradiated: Figure 16a. Extensive bacterial deposits on anterior teeth,
16-year-old male, 2 weeks post irradiation, no caries-preventive measures.   Figure 16b. After 3-1/2 years on preventive
program the teeth remain intact in spite of initial labial decalcification.   Figure 17. Fifty-two-year-old male with no new
caries after 4 years on preventive program.   Exposed cervical
dentin is hard and glossy.   Nonirradiated: Figure 18. Elaborate
bridge work in 56-year-old female suffering from Sjøgren's
syndrome and severe zerostomia.   Figure 19. Posterior teeth
devoid of enamel in 25-year-old male with amelogenesis imperfecta.   Figure 20. Thirty-six-year-old female having suffered
from high caries incidence over many years.   During 2 years
on preventive program one of these three patients (Figure 20)
developed one lesion.

preventive program following radiation developed no carious lesions.

Typical results are illustrated for nonirradiated patients with long histories of rampant caries associated with Sjøgren's syndrome and surgical removal of salivary glands (Figure 18); and for amelogenesis imperfecta (Figure 19) and recurrent caries (Figure 20), both with normal salivary flow. One of these three patients developed one new cavity, the others have remained free of caries during the study period which exceeds two years.

The amount of fluoride deposited in the teeth by the topical procedures is shown in Tables 6 and 7. With regard to penetration and retention of fluoride in enamel (Table 6), it is evident that the levels are considerably elevated as compared to those of untreated teeth (Table 2) and are in the range of fluorosed and carious teeth (Tables 2 and 1). Dentin and cementum (Table 7) show considerably higher levels of fluoride than does enamel. This difference probably reflects the greater permeability of dentin and cementum as well as the smaller size of their apatite crystallites. Arrested carious dentin showed the highest fluoride values of any of the tissues analyzed.

The urine analysis results are shown in Figure 21. It is apparent that most of the patients followed conscientiously the instructions not to swallow the fluoride gel and to rinse the mouth after treatment, as most urine samples showed increases in fluoride excretion of less than 2 mg per 24 hours. However, there were occasional patients who apparently swallowed enough fluoride from a single treatment to increase the 24-hour excretion by 7 mg. This increase in excretion indicates that the amount ingested must have been about 15 mg over the normal level (corresponding to ingestion of 1-1/2 gram of gel).

Ultrastructurally, the arrested carious tissues from treated patient teeth differed from the ordinary carious tissues in several respects. The treated teeth showed evidence of mineral deposition in the form of globular bodies on enamel crystallites (Figure 23), apparently from the surface of the lesion. These mineral deposits resemble those observed in early precipitates (1-2 hours) from the remineralizing solution (Figure 22). Crystallites from subsurface regions of the lesion showed high electron-density and regular outlines indicating apposition of mineral concurrent with the arrest of the lesion. These crystallites resemble more closely those of normal than those of carious enamel (see Johansen, 1965). Sections of embedded enamel

Table 6. Fluoride Content (ppm) of Enamel of Patient Teeth
          Treated In Vivo with Fluoride Gel

Treatment	depth* (μm)					
	1	5	10	25	50	100
APF	3920**	2540	2080	1880	1070	477
	±1480	±1370	±1360	±1450	±985	±412
	13	13	13	11	10	9
NaF	2090	1500	1120	717	425	208
	±723	±729	±497	±365	±304	±119
	12	17	17	17	16	13
APF + NaF	3650	2140	1570	1100	735	338
	±1090	±1147	±647	±604	±595	±329
	12	12	12	12	11	9

*For details see Table 2
**Mean, SD and N.   (T. Olsen and E. Johansen, unpublished)

indicated that the remaining mineral was fused, forming a
honeycomb pattern around enclosed spaces (Figures 25a, b).
At the higher magnification the changes in crystallite morphology resembled those of nonradiation-related caries
(Johansen, 1965). However, fusion of adjacent crystallites
(Figure 25b) was not observed in carious or sound enamel of
untreated teeth.

     Rehardened carious dentin shows a high mineral content
(Figure 26) quite similar to that of sound dentin. While
plate-like crystallites similar to those of the sound and
carious tissues are present, a unique crystal species in the
form of long slender rods is seen within the matrix (Figure
26a) and as sectioned fragments in canal-like spaces (Figure
26b). These crystallites resemble some of those observed in
dentin treated in vitro with remineralizing solution (Figure
9).

     Also, in arrested carious lesions of cementum were cry-
stallites which closely resembled those observed in precipitates from the remineralizing solution (Figure 10). Both
dagger-shaped crystallites (Figure 27a) and long rods
(Figures 27b, c) were commonly seen.

Table 7. Fluoride Content (ppm) of Sound, Exposed and Carious Dentin and Cementum of Teeth Treated In Vivo with Fluoride Gel

Tissue	Layer[+]	Fluoride Gel[++] Treated Teeth Mean SD	Nö.	Non-Treated[+++] Teeth Mean SD	No.
Exposed Dentin	1	11000 ±6200	(3)	227 ±96	(11)
Underlying Dentin	2	2900 ±1300	(3)	281 ±123	(11)
"	3	2700 ±1900	(3)	1160 ±158	(11)
Carious Dentin	1	16600 ±7500	(11)	4900 ±1800	(5)
"	2	11600 ±9400	(10)	1480 ±680	(5)
"	3	6630 ±7280	(9)	950 ±530	(5)
"	4	1900 ±3500	(6)	480 ±410	(5)
Underlying, Normal** Dentin	5	550 ±380	(5)		
Bulk Dentin		430 ±270	(5)	280 ±200	(11)
Exposed Cementum	1	9300 ±4800	(5)	2200 ±510	(10)
Underlying Cementum & Dentin	2	3200 ±2800	(7)	730 ±410	(10)
Underlying, Dentin	3	1300 ±990	(7)	290 ±130	(10)
Calculus		11200 ±4900	(7)	880 ±320	(10)

*Acidulated phosphate fluoride (1.23% F) and/or NaF (1% F).
[+] The combined depth of the 3 layers of exposed dentin was about 1mm. The sampling depth in carious tissue depended on lesion size. The first 3 layers contained visibly stained material, the 4th layer was not stained and the 5th layer** was taken deep below the lesion.
[++] Olsen and Johansen, unpubl. [+++] Olsen and Johansen, 1970,'73.

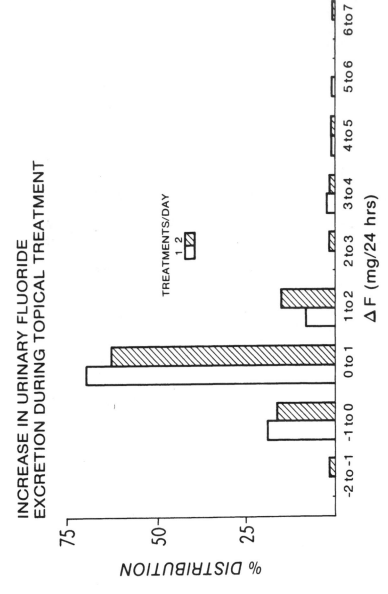

Figure 21. Distribution of urinary fluoride excretion values for patients using fluoride gel once and twice daily. The graph is based on calculated increases in fluoride excretion during gel use over excretion values obtained from the same patient prior to initiation of preventive procedures.

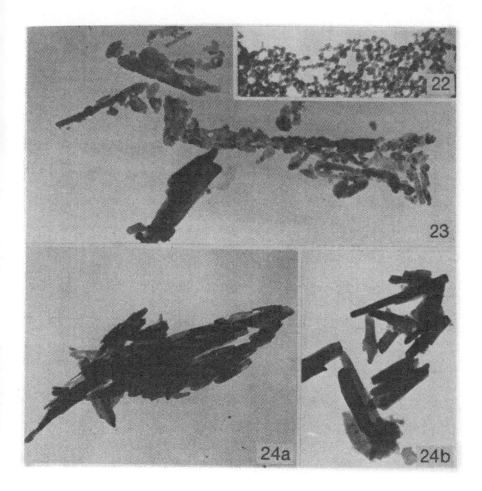

Human Enamel.   EM of enamel from patient on preventive program.
Figure 23.   Triturated sample of arrested carious lesion from
patient started on preventive program with active lesions shows
globular precipitate along enamel crystallites (X 30,000).
These globular bodies are similar to those observed in the precipitate of the remineralizing solution (Figure 22) (X 30,000).
Figure 24a.   Individual crystallites from same lesion as Figure
23 show regular outlines and electron dense structures indicative of remineralization (X 28,000).

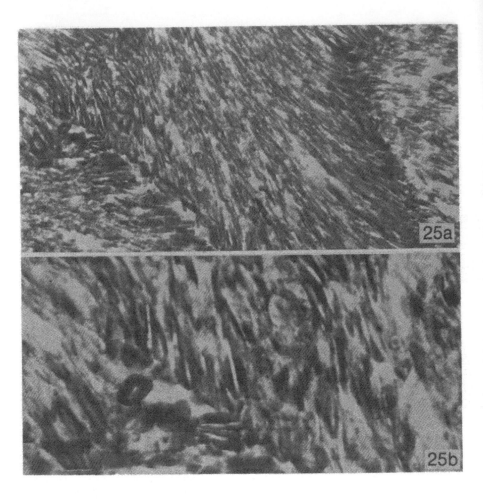

Human Enamel. EM of section of arrested carious enamel from patient on treatment program. The lesion remained arrested for 13
months following initiation of preventive program. Figure 25a.
Enamel rods are clearly evident although the tissue has been extensively demineralized. It appears as if the remaining mineral
has been fused together into a continuous scaffolding structure
embodying spaces most likely containing organic matter
(X 30,000). Figure 25b. At higher magnification individual
crystallites display hollow centers, but rather thick outer
shells. Points of fusion of adjacent crystallites are evident
(X 80,000).

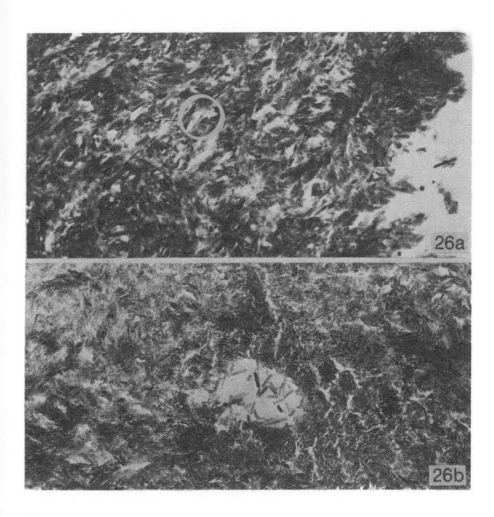

Human Dentin. EM of section of dentin from arrested carious lesion in patient on treatment program. Figure 26a. Long, narrow
rod-shaped crystallites are seen within the rehardened dentin.
These crystallites resembled some of those observed on dentin
treated in vitro with remineralizing solution (X 35,000) (Figure
11). Figure 26b. Sectioned segments of apparently longer rod-
shaped structures dislodged into lumen of dentinal canal. The
length of these fragments reflect the thickness of the section
(X 30,000).

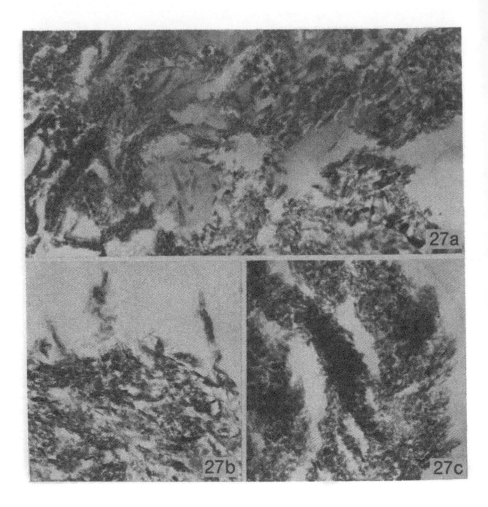

Human Cementum. EM of sections of cementum from arrested lesion
in patient on preventive program. Figure 27a. Elongated, narrow, rod-shaped and dagger-shaped crystallites are seen in broad
surface view (X 20,000). Figure 27b. Elongated rod-shaped
crystallites are intermingled with the smaller plate-like structures making up the bulk of cementum (X 30,000). Figure 27c.
Apparent cross-sectional view of groups of elongated crystallites appearing as punctate images (X 30,000).

Discussion

Daly and Drane (1972) stated that topical treatments with high concentrations of fluoride have to be continued indefinitely to maintain protection against caries in xerostomia patients.* The results reported here, however, clearly show that these treatments can be discontinued after just four weeks. There was no problem in subsequently maintaining most patients (for periods up to four years)on a regimen consisting of oral hygiene procedures, including the use of fluoride toothpaste; daily use of a supersaturated remineralizing mouthwash; and salivary stimulation by chewing of gum.

Numerically, comparison of the various studies on xerostomia patients is difficult for several reasons. Because of the severe consequences of abscessed teeth in irradiated patients, only two studies (Daly and Drane, 1972; Dreizen et al., 1976) have used randomized controls. In our study the controls were partly self-selected and partly selected by circumstance (i.e., irradiation before preventive program started, or failure of referral system). Only in one previous report (Dreizen et al.) are time-related rates of decay reported; in most cases only the total length of time of the study is given, not the average time that the patients were followed. Also, most studies preselected patients and/ or teeth, i.e., teeth or patients deemed unlikely to respond successfully to the regimen were eliminated from the program. For example, Keyes and McCasland (1976) report extracting 22% of their patients' teeth before irradiation. No such preselection was used in our study; no teeth were extracted initially or during the study because of caries.

In our study, the treatment group had 0.2, 0.2 and 0.3 new lesions per patient per year for the first, second and third year, respectively. These results compare favorably with those of Dreizen et al. for the group using fluoride gel on a daily basis. That group had caries increments of 0.07 DMFT per month (or 0.84 DMFT per year). It should be noted that in our system, each new lesion was counted, while in the DMFT system a tooth is only "at risk" if it has no previous decay or restoration. Thus the number of lesions would normally be higher than the DMFT count if applied to the same group of patients.

No information on the amount of fluoride ingested from the topical treatments have been reported from the studies

---

*This view was also expressed to us by Dr. S. Dreizen during
 a U.S.P.H.S. project site visit on November 20, 1975 to
 review this project.

on irradiated patients. However, our data on urinary fluoride excretion of patients using topical gels show that a
few patients will, in spite of instructions to the contrary,
swallow enough fluoride to be of concern if the ingestion
were continued for long periods of time (see Taves; and
Jowsey et al., this monograph). A program of indefinite use
of daily gel applications, therefore, should not be recommended without monitoring to prevent chronic effects.

The contributions of each step of the preventive system
has not yet been adequately assessed, but interaction between
the various procedures, or superadditive effects, are likely
to have occurred. The reason for the difference between our
findings and those of Daly and Drane, and Dreizen et al. may
be that in our study each topical gel treatment was followed
by a remineralizing mouthwash. The solution creates conditions conducive to nucleation and mineralization that would
be expected to help incorporate into the apatite structure
fluoride which was originally surface adsorbed or deposited
as CaF_2. In the dynamic environment of the tooth, such incorporation would reduce subsequent loss of fluoride due to
dissolution and ionic exchange processes.

The continued use of fluoride toothpaste followed by
remineralizing solution was intended to maintain the protective fluoride levels created by the four-week use of gel.
Previous studies on exposed dentin and cementum (Furseth and
Johansen, 1968; Olsen and Johansen, 1970, 1973) have shown
that in the normal oral environment an overall force for
deposition of high fluoride mineral exists, even with the
low levels of fluoride normally present in the oral fluid.
This force may be considered part of an inherent cariostatic
mechanism of the oral fluid-apatite system. (It finds expression even under conditions of caries formation where
the amount of fluoride per unit volume of tissue often increases (Levine, 1973; T. Olsen and E. Johansen, unpublished)
and the crystals frequently are larger (Johansen and Parks,
1961; Furseth and Johansen, 1970) than in the sound tissues.
If this physiological cariostatic mechanism does not function adequately, as in patients with severe xerostomia, two
possible courses of remedial action are apparent. The
bacterial attack rate may be decreased by frequent use of an
antibacterial agent (fluoride gel when used daily, as by
Daly and Drane, may have acted primarily this way); or the
physiological maintenance system can be stimulated (chewing
gum) and augmented (artificial mineralizing solution with
low concentrations of fluoride). That the latter approach
provides adequate protection is supported by the observation that the caries rate did not change significantly over
three years (Table 5). Thus, the results suggest that our

premise is correct. Also, the chemical data from extracted
patient teeth show that the fluoride levels in treated enamel
(Table 6) are comparable to those found in mineral persisting
in carious lesions of enamel of untreated teeth, and the
fluoride levels of treated cementum and superficial layers
of exposed dentin far exceeded those found in mineral persisting in carious lesions of these tissues in untreated
teeth (Table 7). Thus the fluoride content of all treated
tissues have reached or exceeded the levels we predicted to
be associated with high caries resistance. Remineralizing
solution when used alone, did not show a clear anticaries
effect in vivo in the only published study (McCormick et al.,
1965, 1970). A second in vivo study (Levine, 1975) did indicate some effectiveness in accelerating salivary rehardening of induced early lesions, but the author stated that the
implications in relation to the clinical situation were unclear. To our knowledge, however, no other study on caries
prevention had, at the time of this symposium, incorporated
the use of supersaturated calcium phosphate solutions as an
adjunct to topical fluoride treatment.

Oral hygiene has long been considered important in the
control of caries, both on a theoretical basis (removal of
acid-producing bacteria) and in the judgement of many dental
practitioners. Attempts to demonstrate a close relationship
by objective study have, however, often led to inconclusive
results (Bibby, 1966; Ripa et al., 1977). The failure to
find a convincing relationship may be due to the experimental
designs which have often involved assessment of frequency of
use of oral hygiene procedures by questioning the patients
rather than representative and objective measurement of
their effectiveness. Also, it has been a frequent observation (e.g., Heifetz, 1973) that most individuals, even after
receiving thorough instruction, have difficulty in cleaning
their teeth adequately with the commonly used brushing and
flossing techniques. This was found to be the case with
many of our patients until the additional technique of
cleaning with cotton swabs was introduced. This method
proved particularly effective in removing debris from cervical areas and from the distal surface of the most posterior
teeth.

Saliva stimulation by drugs such as pilocarpine have
been used in experimental studies. However, the salivary
glands can be stimulated without the use of drugs (i.e., by
chewing), and this seemed a better long-term approach to the
problem of lack of saliva. In our study a chewing gum containing no sugar or sugar substitute was found effective in
increasing the salivary flow and thus enhancing the protective effects of saliva as well as alleviating the discomfort

of a dry mouth. In irradiated patients the best results
seemed to be obtained if use of the gum was started at or
before initiation of radiation therapy.

Concluding Remarks

Review of the literature shows that topical fluorides
as presently used for caries control provide highly variable
protection against the disease. Maximum benefits have only
been achieved by continued daily use of highly concentrated
solutions. Inherent in this procedure is the risk of chronic
ingestion of harmful amounts of fluoride. The data presented
here show that four weeks' use of fluoride gel is sufficient
when it is part of a multifaceted preventive program designed to increase tooth resistance, decrease the acid attack
and enhance the physiological maintenance processes. This
program has been successfully applied to various groups of
highly susceptible individuals, thus confirming the validity
of the multifaceted approach. In application to the normal
population, the practitioner may want to modify the program
to meet the needs of the individual patient. However,
standardized procedures similar to those used in this study
could be expected to provide sufficient protection to allow
for less than ideal compliance and hence would be suitable
for mass application.

Acknowledgement

Financial support for some aspects of these studies were received from the following sources:

USPHS National Institute of Dental Research
    DE 00689 (terminated 1975)

USPHS National Cancer Institute CA 11198

Highland Hospital, Rochester, N.Y.

University of Rochester

References

[1]*Averill, H.M., Averill, J.E. and Ritz, A.G. (1967) A two-
    year comparison of three topical agents. J.Am.dent.
    Assoc., 74; 996-1001.

---

*Numbers above references refer to tabulations in Figures
1 and 2.

Berggren, H. (1967) Topical fluorides (including dentifrices). Int.Dent.J., 17; 40-46.

Berggren, H. and Welander, E. (1960) Supervised toothbrushing with a sodium fluoride solution in 5000 Swedish school children. Acta odont.Scand., 18; 209-34.

[2]Bergman, G. (1953) Topical application of sodium fluoride using school children as subjects. Acta odont.Scand., 11, suppl., 12; 53-112.

Bibby, B.G. (1946) Topical applications of fluorides as a method of combating dental caries. In Dental Caries and Fluorine, (F.A. Moulton, ed.), AAAS, Washington, pp. 93-98.

Bibby, B.G. (1966) Do we tell the truth about preventing dental caries? J.Dent.Child., 33; 269-79.

Bibby, B.G. and Van Kestern, M. (1940) The effect of fluorine on mouth bacteria. Odont.Res., 19; 391-402.

[3]Bixler, D. and Muhler, J.C. (1962) Experimental clinical human caries test design and interpretation. J.Am.dent.Assoc., 65; 482-488.

[4]Bixler, D. and Muhler, J.C. (1966) Effectiveness of a stannous fluoride-containing dentifrice in reducing dental caries in children in a boarding school environment. J.Am.dent.Assoc., 72; 653-658.

Broukal, Z. and Zajicek, O. (1974) Amount and distribution of extracellular polysaccharides in dental microbial plaque. Caries Res., 8; 97-104.

Brown, W.E., Gregory, T.M. and Chow, L.C. (1977) Effects of fluoride on enamel solubility and cariostasis. In Cariostatic Mechanisms of Fluorides. (W.E. Brown and K.G. König, eds.), Caries Res., 11 (Suppl.1); 118-35.

Brown, W.E. and König, K.G. (Editors) (1977) Cariostatic Mechanisms of Fluorides, Caries Res., 11, suppl. 1.

Brudevold, F. and McCann, H.G. (1968) Enamel solubility tests and their significance in regard to dental caries. Ann.N.Y.Acad.Sci., 153(1); 20-51.

Brudevold, F., Gardner, D.E. and Smith, F.A. (1956) The distribution of fluoride in human enamel. J.dent.Res., 35; 420-29.

Brudevold, F., Savory, A. and Bowman, L. (1957) The effect of fluoride in phosphate buffers on enamel. IADR Abstr. #72.

Brudevold, F., Steadman, L.T. and Smith, F.A. (1960) Inorganic and organic components of tooth structure. Ann. N.Y.Acad.Sci., 85; 110-32.

Brudevold, F. et al. (1963) A study of acidulated fluoride solutions. I. In vitro effects on enamel. Archs.oral Biol., 8; 167-177.

Brudevold, F. et al. (1975) Determination of trace elements in surface enamel of human teeth by a new biopsy procedure. Archs.oral Biol., 20; 667-673.

Brudevold, F. et al. (1967) The chemistry of caries inhibition. Problems and challenges in topical treatments. J.dent.Res., 46; 37-45.

[5]Bryan, E.T. and Williams, J.E. (1970) The cariostatic effectiveness of a phosphate fluoride gel administered annually to school children, final results. J.Publ. Hlth.Dent., 30; 13-66.

Bullen, D.C.T., McCombie, F. and Hole, L.W. (1966) Two year effect of supervised toothbrushing with an acidulated fluoride-phosphate solution. J.Can.dent.Assoc., 32; 89-93.

Carl, W., Schaaf, N.G. and Chen, T.Y. (1972) Dental care of patients irradiated for cancer of the head and neck. Cancer, 30; 448-53.

[6]Cartwright, H.V., Lindahl, R.L. and Bawden, J.W. (1968) Clinical findings on the effectiveness of stannous fluoride and acid phosphate as caries-reducing agents in children. J.dent.Child., 35; 36-40.

Cloutier, P.F. and Johansen, E. (1977) Discrepancies between drinking water and urinary fluoride. AADR Abstr. #521.

[7]Cons, N.C., Janerich, D.T. and Senning, R.S. (1970) Albany topical fluoride study. J.Am.dent.Assoc., 80; 777-781.

Cutress, T.W. (1972) The inorganic composition and solubility of dental enamel from several specified population groups. Archs.oral Biol., 17; 93-109.

Daly, T.E. (1973) Dental care in the irradiated patients. In Textbook of Radiology, Ed. Fletcher, G.H., Lea and Febiger 2nd Edn, p. 157.

Daly, T.E. and Drane, J.B. (1972) The management of teeth related to the treatment of oral cancer. In Proceedings 7th National Cancer Conference, J.B. Lippincott Co., 1973.

Daly, T.E., Drane, J.B. and MacComb, W.S. (1972) Management of problems of the teeth and jaw in patients undergoing irradiation. Am.J.Surg., 124; 539-42.

Darling, A.E. (1963) Microstructural changes in early dental caries. In Mechanisms of Hard Tissue Destruction, Ed. Sognnaes, R.F., Publ. #75, AAAS, Washington, D.C.

Davies, G.N. (1974) Cost and Benefit of Fluoride in the Prevention of Dental Caries. WHO, Geneva, 1974, p. 27.

Downs, R.A. and Pelton, W.J. (1950) The effect of topically applied fluorides in dental caries experience on children residing in fluoride areas. J.Colo.dent.Assoc., 29; 7-10.

Dreizen, S. et al. (1976) Prevention of xerostomia - induced caries in irradiated cancer patients. IADR Abstr. #92.

[8]Englander, H.R. et al. (1969) Residual anti-caries effect of repeated topical applications by mouthpieces. J.Am. dent.Assoc., 78; 738-787.

[9]Englander, H.R. et al. (1967) Clinical anti-caries effect of repeated topical sodium fluoride applications by mouthpieces. J.Am.dent.Assoc., 75; 638-644.

Englander, H.R. et al. (1971) Incremental rates of dental caries after repeated topical sodium fluoride applications in children with life-long consumption of fluoridated water. J.Am.dent.Assoc., 82; 354-58.

Ericson, T. and Ericsson, Y. (1967) Effect of partial fluorine substitution on the phosphate exchange and protein adsorption of hydroxyapatite. Helv.odont.Acta, 11; 10-11.

[10]Fanning, E.A., Gotjamanos, T. and Rowles, N.J. (1968) The use of fluoride dentifrices in the control of dental caries: Methodology and results of a clinical trial. Aust.dent.J., 13; 201-206.

Forrester, D.J. and Schulz, E.M. (Editors) (1974) International Workshop on Fluorides and Dental Caries Reductions. University of Maryland School of Dentistry, Baltimore.

Furseth, R. and Johansen, E. (1968) A microradiographic comparison of sound and carious human dental cementum. Archs.oral Biol., 13; 1197-1206.

Furseth, R. and Johansen, E. (1970) The mineral phase of sound and carious human dental cementum studied by electron microscopy. Acta odont.Scand., 28; 305-322.

[11]Galagan, D.J. and Knutson, J.W. (1947) The effect of topically applied fluorides on dental caries experience. V. Report of findings with two, four and six applications of sodium. fluoride and lead fluoride. Publ.Hlth. Rep., 62; 1477-1483.

[12]Galagan, D.J. and Vermillion, J.R. (1955) Effect of topical fluorides on teeth matured on fluoride-bearing water. Publ.Hlth.Rep., 70; 1114-1115.

[13]Gish, C.W. and Muhler, J.C. (1965) Effectiveness of a SnF_2-$Ca_2P_2O_7$ dentifrice on dental caries of children whose teeth calcified in a natural fluoride area. I. Results at the end of 12 months. J.Am.dent.Assoc., 73; 853-855.

[14]Idem. (1966) II. Results at the end of 24 months. J.Am. dent.Assoc., 73; 853-855.

Glantz, P.O. (1969) Wettability and Adhesiveness. Odont. Revy., 17; Suppl., pp. 1-132.

Glas, J.E., Nylen, M.V. and Little, M.F. (1965) A microradiographic and electron microscopic study of carious enamel. IADR Abstr. #121.

Gray, J.A., Francis, M.D. and Griebstein, W.J. (1962) Chemistry of enamel dissolution. In Chemistry and Prevention of Dental Caries, Ed. R.F. Sognnaes, pp.164-179, Charles C. Thomas, Springfield, Ill.

[15]Harris, H. (1963) Observations on the effect of eight percent stannous fluoride on dental caries in children.
Austr.dent.J., 8; 335-340.

Heifetz, S.B. (1973) Programs for the mass control of
plaque; an appraisal. J.Publ.Hlth.Dent., 33; 91-95.

Hiatt, W.H. and Johansen, E. (1972) Root preparation. 1.
Obturation of dentinal tubules in treatment of root
hypersensitivity. J.Periodont., 43; 373-380.

[16]Horowitz, H.S. (1968) The effect on dental caries of topically applied acidulated phosphate-fluoride: Results
after one year. J.oral Ther., 4; 286-291.

[17]Horowitz, H.S. (1969) Effect on dental caries of topically
applied acidulated phosphate-fluoride: Results after
two years. J.Am.dent.Assoc., 78; 568-572.

[18]Horowitz, H.S., Creighton, W.E. and McClendon, B.J. (1971)
The effect on human dental caries of weekly oral rinsing with a sodium fluoride mouthwash. Archs.oral Biol.,
16; 609-616.

[19]Horowitz, H.S. and Doyle, J. (1971) The effect on dental
caries of topically applied acidulated phosphate-fluoride. Results after 3 years. J.Am.dent.Assoc., 82;
359-365.

Horowitz, H.S. and Heifetz, S.B. (1968) A review of studies
on the self-administration of topical fluorides. Can.
J.Publ.Hlth., 59; 393-98.

Horowitz, H.S. and Heifetz, S.B. (1969) Evaluation of topical applications of stannous fluoride to teeth of children born and reared in a fluoridated community. Final
report. J.dent.Child., 36; 65-71.

[20]Horowitz, H.S. et al. (1966) Evaluation of a stannous fluoride dentifrice for use in dental public health programs. J.Am.dent.Assoc., 72; 408-422.

[21]Horowitz, H.S. and Lucye, H.S. (1966) A clinical study of
stannous fluoride in a prophylaxis paste and as a solution. J.oral Ther.Pharm., 3; 17-25.

[22]Howell, C.L. et al. (1955) Effect of topically applied
stannous fluoride on dental caries experience in children. J.Am.dent.Assoc., 50; 14-17.

[23] Ingraham, R.Q. and Williams, J.E. (1970) An evaluation of the utility of application and cariostatic effectiveness of phosphate-fluorides in solution and gel states. J. Tenn.dent.Assoc., 50; 5-12.

Isaac, S. et al. (1958) Solubility rate and natural fluoride content of surface and subsurface enamel. J.dent.Res., 37; 254-263.

[24] Jakovljevic, D. (1972) Research in the use of fluorides with Zagreb children, Copenhagen, WHO Regional Office for Europe (document EURO-5506). Cited from Davies (1974).

Johansen, E. (1962) The nature of the carious lesion. In The Dental Clinics of North America, Consult. ed., W.B. Saunders Co., Philadelphia, July, p. 305-320.

Johansen, E. (1963) Ultrastructural and chemical observations on dental caries. In Mechanisms of Hard Tissue Destruction, R.F. Sognnaes, ed., AAAS, Washington, D.C. p. 187-211.

Johansen, E. (1964) Microstructure of enamel and dentin. J.dent.Res., 43; 1007-1020.

Johansen, E. (1965) Comparison of the ultrastructure and chemical composition of sound and carious enamel from human permanent teeth. In Tooth Enamel, M.V. Stack and R.W. Fearnhead, eds., Williams & Wilkins, Baltimore, p. 177-181.

Johansen, E. (1967) Ultrastructure of dentin. In Structural and Chemical Organization of Teeth, A.E.W. Miles, ed., Academic Press, London, pp. 35-73.

Johansen, E. (1969) The ultrastructure and chemistry of carious lesions. Proc., First Pan Pacific Congress on Dent.Res., pp. 31-38.

Johansen, E. and Nordback, L.G. (1962) The chemistry of carious lesions III. The fluoride content of carious dentin. IADR Abstr. #143.

Johansen, E. and Parks, H.F. (1961) Electron microscopic observations on soft carious dentin. J.dent.Res., 40; 235-248.

Johansen, E. et al. (1975) Mechanisms of radiation caries and dental management of irradiated patients. Presentation and workshop - American Society of Therapeutic Radiologists, Annual Meeting, October 8-12, 1975, San Francisco, Calif.

[25]Jordan, W.A. and Peterson, J.K. (1959) Caries-inhibiting value of a dentifrice containing stannous fluoride: Final report of a two-year study. J.Am.dent.Assoc., 58; 42-44.

[26]Keyes, F.M., Overton, N.J. and McKean, T.W. (1961) Clinical trials of caries inhibitory dentifrices. J.Am.dent. Assoc., 63; 189-193.

Keyes, H.M. and McCasland, J.P. (1976) Techniques and results of a comprehensive dental care program in head and neck cancer patients. Int.J.Rad.Oncol.Biol.Phys., 1; 859-65.

Kleinberg, I. (1970) Biochemistry of the dental plaque. In Advances in Oral Biology, Staple, P.H., Ed., Vol. 1, pp. 44-90.

Knutson, J.W. and Armstrong, W.D. (1943) Effect of topically applied sodium fluoride on dental caries experience. Publ.Hlth.Rep., 58; 1701-1715.

Koch, G. (1967) Effect of sodium fluoride in dentifrice and mouthwash on incidence of dental caries in school children. Odont.Revy.18 (Suppl.12).

[27]Koch, G. (1969) Caries increments in school children after the end of supervised rinsing with sodium fluoride solution. Odont.Revy., 20; 323-330.

Koch, G. and Petersson, L.G. (1975) Caries preventive effect of a fluoride-containing varnish (Duraphat) after 1 year's study. Comm.dent.Oral Epidemiol., 3; 262-66.

Koulourides, T., Feagin, F. and Pigman, W. (1965) Rehardening of dental enamel by saliva in vitro. In Mechanisms of Dental Caries, Ann.N.Y.Acad.Sci., 131(2); 771-775.

Koulourides, T., Feagin, F. and Pigman, W. (1968) Effect of pH, ionic strength, and cupric ions on the rehardening rate of buffer softened enamel. Archs.oral Biol., 13; 335-41.

[28]Law, F.E., Jeffreys, M.H. and Sheary, H.C. (1961) Topical applications of fluoride solutions in dental caries control. Publ.Hlth.Rep., 76; 287-290.

Levine, R.S. (1973) The differential inorganic composition of dentine within active and arrested carious lesions. Caries Res., 7; 245-60.

Levine, R.S. (1975) An initial clinical assessment of a mineralizing mouthrinse. Brit.dent.J., 138; 249-52.

Marquis, R.D., Lisher, R.J. and Bronsten, R. (1976) Growth of plaque bacteria in pellet culture and the lytic effect of fluoride. Proc.Microbiol. Aspects of Dental Caries. Eds. Stiles, Loesche and O'Brien, Vol. III; 821-28.

Marthaler, T.M. (1968) Caries-inhibition after seven years of unsupervised use of an amine fluoride dentifrice. Brit.dent.J., 124; 510-15.

Marthaler, T.M., König, K.G. and Mühlemann, H.R. (1970) The effect of fluoride gel used for supervised tooth-brushing 15 to 30 times per year. Helv.odont.Acta, 14; 67-77.

McCann, H.G. (1968) Inorganic components of salivary secretions. In Art and Science of Dental Caries Research, Harris, R.S. (ed.), Academic Press, N.Y.; London.

McCormick, J. and Koulourides, T. (1965) A study of neutral calcium, phosphate, and fluoride remineralizing mouthwashes. IADR Abstr. #402.

McCormick, J. et al. (1970) Remineralizing mouthwash; rationale and a pilot clinical study. Ala.J.Med.Sci., 7; 92-97.

[29]Mercer, V.H. and Muhler, J.C. (1961) Comparison of a single application of stannous fluoride with a single application of sodium fluoride or two applications of stannous fluoride. J.dent.Child., 28; 84-86.

[30]Mercer, V.H. and Muhler, J.C. (1964) The effect of 30 second topical stannous fluoride on dental caries reductions in children. J.oral.Ther.Pharm., 1; 141-146.

[31]Mergele, M. (1968a) Report I. A supervised brushing study in state institution schools. Bull.Acad.Med. NJ, 14; 247-250.

[32]Mergele, M. (1968b) Report II. An unsupervised brushing study on subjects residing in a community with fluoride in the water. Bull.Acad.Med. NJ, 14; 251-255.

[33]Møller, I.J., Holst, J.J. and Sørensen, E. (1968) Caries-reducing effect of a sodium monofluorophosphate dentifrice. Brit.dent.J., 124; 209-213.

[34]Muhler, J.C. (1958) Topical treatment of the teeth with stannous fluoride. Single application technique. J. dent.Child., 25; 306-309.

[35]Muhler, J.C. (1959) Present status of topical fluoride therapy. J.dent.Child., 26; 173-185.

[36]Muhler, J.C. (1962) Effect of a stannous fluoride dentifrice on caries reduction in children during a three-year study period. J.Am.dent.Assoc., 64; 216-224.

[37]Muhler, J.C. (1970) A clinical comparison of fluoride and antienzyme dentifrices. J.dent.Child., 37; 501-502, passim 511-514.

[38]Muhler, J.C. and Radike, A.W. (1957) Effect of a dentifrice containing stannous fluoride on dental caries in adults. II. Results at the end of two years of unsupervised use. J.Am.dent.Assoc., 55; 196-198.

[39]Muhler, J.C. et al. (1954) The effect of a stannous fluoride-containing dentifrice on caries reduction in children. J.dent.Res., 33; 606-612.

[40]Muhler, J.C. et al. (1955a) Effect of a stannous fluoride-containing dentifrice on caries reduction in children. II. Caries experience after one year. J.Am.dent.Assoc., 50; 163-166.

[41]Muhler, J.C. et al. (1955b) A comparison between the anti-cariogenic effects of dentifrices containing stannous fluoride and sodium fluoride. J.Am.dent.Assoc., 51; 556-559.

[42]Muhler, J.C. et al. (1956) The effect of a stannous fluoride-containing dentifrice on dental caries in adults. J.dent.Res., 35; 49-53.

Murray, J.J., Winter, G.B. and Hurst, C.P. (1977) Duraphat fluoride varnish. A 2-year clinical trial in 5-year-old children. Brit.dent.J., 143; 11-17.

[43]Naylor, M.N. and Emslie, R.D. (1967) Clinical testing of stannous fluoride and sodium monofluorophosphate dentifrices in London school children. Brit.dent.J., 123; 17-23.

Neuman, W.F. and Neuman, M.W. (1958) The chemical dynamics of bone mineral. University of Chicago Press, Chicago.

Olsen, T. and Johansen, E. (1970) Inorganic composition of sound unaltered, exposed, sclerosed and carious dentin. IADR Abstr. #46.

Olsen, T. and Johansen, E. (1972) Inorganic composition of sound and carious human cementum. IADR Abstr. #174.

Olsen, T. and Johansen, E. (1973) Inorganic composition of exposed noncarious cementum. IADR Abstr. #274.

Olsen, T. and Johansen, E. (1977) In vitro comparison of topical fluoride preparations and application procedures. AADR Abstr. #634.

[44]Peffley, G.E. and Muhler, J.C. (1960) The effect of a commercial stannous fluoride dentifrice under controlled brushing habits on dental caries incidence in children; Preliminary report. J.dent.Res., 39; 871-874.

[45]Peterson, J.K. and Williamson, L. (1962) Effectiveness of topical application of eight percent stannous fluoride. Publ.Hlth.Rep., Washington, 77; 39-40.

Pigman, W., Cueto, H. and Baugh, D. (1964) Conditions affecting rehardening of softened enamel. J.dent.Res., 43; 1187-1195.

Reed, A.J. and Bibby, B.G. (1976) Preliminary report on effect of topical application of titanium tetrafluoride on dental caries. J.dent.Res., 55; 357-58.

Regezi, J.A., Courtney, R.M. and Kerr, D.A. (1976) Dental management of patients irradiated for oral cancer. Cancer, 38; 994-1000.

Ripa, L.W., Barenie, J.T. and Liske, G.S. (1977) The relationship between oral hygiene and dental health. An epidimological survey. NY State Dent.J., 43; 530-35.

[46]Rugg-Gunn, A.J., Holloway, P.J. and Davies, T.G.H. (1973) Caries prevention by daily fluoride mouthrinsing. Brit. dent.J., 135; 353-360.

[47]Salter, W.A.T., McCombie, F. and Hole, L.W. (1962) The anti-cariogenic effects of one and two applications of stannous fluoride on the deciduous and permanent teeth of children age 6 and 7. J.Can.dent.Assoc., 28; 363-371.

Singh, B. (1961) Chemical studies on sound and carious human enamel. MS Thesis, University of Rochester.

[48]Szwedja, J.F. (1971) Fluorides in community programs: Results after four years of study of various agents topically applied by two technics. J.Publ.Hlth.Dent., 31; 166-176.

Tinanoff, N., Brady, J.M. and Gross, A. (1976) The effect of NaF and SnF mouthrinses on bacterial colonization of tooth enamel. TEM and SEM studies. Caries Res., 10; 415-26.

[49]Thomas, A.E. and Jamison, H.C. (1966) Effect of SnF dentifrices on caries in children: Two-year clinical study of supervised brushing in children's homes. J.Am.dent. Assoc., 73; 844-852.

[50]Thomas, A. and Jamison, H. (1970) Effect of a combination of two cariostatic agents in children: Two-year clinical study of supervised brushing in children's homes. J.Am.dent.Assoc., 81; 118-124.

[51]Torell, P. and Ericsson, Y. (1965) Two year clinical tests with different methods of local caries preventive fluoride applications in Swedish school children. Acta odont.Scand., 23; 287-322.

Volker, J.F. (1939) Effect of fluorine on solubility of enamel and dentin. Proc.Soc.Exp.Biol.and Med., 42; 725-27.

Volker, J.F. and Bibby, B.G. (1941) The action of fluorine in limiting dental caries. Medicine, 20; 211-27.

Weber, J.C., Posner, A.S. and Eaves, E.D. (1967) Morphology of synthetic and biological amorphous calcium phosphates. IADR Abstr. #73.

Weiss, S. et al. (1965) Influence of various factors on polysaccharide synthesis in S.mitis. Ann.N.Y.Acad.Sci., 131; 839-50.

[52]Wellock, W.D. and Brudevold, F. (1963) A study of acidulated fluoride solutions. II. The caries inhibiting
effect of single annual topical applications of an
acidic fluoride and phosphate solution. A two-year
experience. Archs.oral Biol., 8; 179-182.

[53]Wellock, W.D., Maitland, A. and Brudevold, F. (1965) Caries
increments, tooth discolouration and state of oral
hygiene in children given single applications of acid
phosphate - fluoride and stannous fluoride. Archs.oral
Biol., 10; 453-460.

Wescott, W.B., Starcke, E.N. and Shannon, I.L. (1975)
Chemical protection against postirradiation dental
caries. Oral Surg., Oral Med., Oral Path., 40; 709-
719.

Wright, D.E. and Jenkins, G.N. (1954) The effect of fluoride on the acid production of saliva glucose mixtures.
Brit.dent.J., 96; 30-33.

[54]Zacherl, W.A. and McPhail, C.W.B. (1965) Evaluation of a
stannous fluoride-calcium pyrophosphate dentifrice.
J.Can.dent.Assoc., 31; 174-180.

[55]Zacherl, W.A. and McPhail, C.W.B. (1970) Final report on
the efficacy of a stannous fluoride-calcium pyrophosphate dentifrice. J.Can.dent.Assoc., 36; 262-264.

Fluoride in the Treatment of Osteoporosis

Jenifer Jowsey, B.L. Riggs and P.J. Kelly

Doses of fluoride effective in the treatment of osteoporosis involve levels of fluoride that are approximately ten times higher than those which have been previously discussed in this symposium and in papers concerned with the prevention of tooth decay. The mechanism of action of fluoride on the mineralized tissue, in this instance bone, is also different from those in the prevention of dental caries: It is one of osteoblastic stimulation, rather than the prevention of crystal dissolution (Jowsey et al., 1972).

Clinical diagnosis of osteoporosis is generally based on the presence of a symptomatic fracture because this provides the clearest definitions of the disease. Our investigations have included patients with spinal fractures and have excluded patients with Colles' fractures (fracture of the distal forearm) and femoral neck; although the latter fractures are also complications of osteoporosis. The following report will outline briefly the major characteristics of osteoporosis and describe our investigations into the use of fluoride in the treatment of the disease.

A radiograph of the spine of a patient with symptomatic osteoporosis will show wedging or anterior collapse of the vertebrae which results in kyphosis and symptomatic back pain. Symmetrical collapse will result in the loss of height which is also a characteristic feature of the osteoporotic individual. The fracture or fractures occur because there is an insufficient amount of bone to sustain normal stress. The patients have symptoms as the result of fractures and on this basis are separable from asymptomatic individuals with a similarly decreased amount of bone but who have not sustained a fracture and therefore by the definition of the disease do not fall in the category of osteoporosis.

In osteoporosis the bone is histologically normal, and

Table 1. Bone Resorption and Formation in Control and
Osteoporotic Populations.

Bone Activity	N	Mean	S.D.
Formation			
30-50 yr. controls	28	2.1	1.1
30-50 yr. osteoporotics	18	2.4	1.7
51-100 yr. controls	62	2.0	1.4
51-100 yr. osteoporotics	144	2.6	0.1
Resorption			
30-50 yr. controls	28	3.5	0.9
30-50 yr. osteoporotics	18	9.4	0.8
51-100 yr. controls	62	4.0	1.5
51-100 yr. osteoporotics	144	10.9	5.2

Values are formation or resorption as percent of total
surface.

Table 2. Bone Mass in the Iliac Crest.

Age	n	% bone* mean ± S.D.
20 - 29	11	42.4 ± 8.2
30 - 39	11	38.8 ± 15.5
40 - 49	13	31.0 ± 8.1
50 - 59	17	32.4 ± 9.0
60 - 60	22	24.8 ± 8.0
70 - 100	24	24.7 ± 9.2

*Percent bone in transilial biopsy sample.

in the United States where vitamin D is added to milk and where exposure to sunshine is either adequate or above average, osteoporosis is only occasionally complicated by osteomalacia. However, this is not true in other parts of the world, particularly northern Europe, where osteomalacia in fact complicates the histological picture relatively frequently (Chalmers et al., 1967).

In bone biopsy specimens from patients with symptomatic fractures of the spine, the most clearly distinguishing feature is an increase in bone resorption surfaces (Table 1). The bone loss resulting in fractures and pain appears to be caused by a negative calcium balance which probably existed in these individuals since young adulthood and which has resulted in mineralized bone being removed from the skeleton in order to maintain a normal serum calcium level (Table 2). In other words the skeleton appears to act as a source of calcium in situations in which there are inadequate supplies of calcium in the diet and when excessive amounts of calcium are lost through the gut, as in malabsorption, or excreted in the urine.

Etiology
In some cases osteoporosis is iatrogenic but with the exception of steroid-induced osteoporosis this is a rare finding (Table 3). Hormonal inbalance may result in osteoporosis, with the disease usually being clearly related to an excessive level of the hormone in the serum. Estrogen deficiency, in the postmenopausal female, also causes acceleration of bone loss and therefore accentuates the disease. However, some women become symptomatically osteoporotic before this deficiency can be expected to have had any significant effect on bone loss. In addition, bone density data suggests that bone loss begins soon after skeletal maturity is reached, at age 20 (Table 2). For early loss of bone, high levels of dietary phosphorus in combination with low calcium intake may be largely to blame (Jowsey et al., 1974). Inactivity also contributes to bone loss and perhaps may account for the age at which osteoporosis presents itself.

Treatment
The discussion to follow will briefly summarize various forms of treatment that have been studied both by us and by a number of other investigators, and will then concentrate primarily on the use of fluoride and calcium in the treatment of the disease.

Calcitonin has been suggested as being useful, based on the theory that it decreases bone resorption. However, evaluation of the many studies in which calcitonin has been used for the treatment of postmenopausal or senile osteoporosis

Table 3.   Etiological Factors in Osteoporosis.

Iatrogenic

Gastrectomy--poor absorption of both calcium and vitamin D.
Anticonvulsant agents--failure of conversion of vitamin D to
1-25-OH vitamin D.
Steroid administration.

Nutritional

Calcium deficiency--lactose intolerance.
Vitamin D deficiency--from diet and environment.
Protein deficiency (Kwashiorkor) or vitamin C deficiency
  (scurvy)*

Hormonal

Hyperthyroidism (thyrotoxicosis)
Hyperparathyroidism
Acromegaly
Hypercortisonism (Cushing's syndrome)
Diabetes mellitus
Estrogen deficiency--postmenopausal

Genetic

Osteogenesis imperfecta
Estrogen deficiency--Turner's syndrome

Associated Soft-Tissue Disease

Malabsorption--sprue
Renal failure
Metabolic acidosis*
Mast cell disease*
Lymphoma

---

*uncommon in the U.S.

has failed to produce convincing data that this hormone will cause an increase in bone mass or cessation of fractures (Jowsey et al., 1971; Melick et al., 1973). Indeed, opinion is divided as to whether bone resorption is decreased, increased or remains the same. Diphosphonates are known to deposit on bone surfaces, and it has been suggested that they could prevent resorption of bone and, therefore, might be therapeutically useful in osteoporosis. However, patients with both osteoporosis and Paget's disease given EHDP, a disodium diphosphonate, develop large quantities of unmineralized osteoid, and their serum parathyroid hormone levels tend to increase (Jowsey et al., 1971a). Oral phosphate supplements were considered useful because they result in a decrease in urinary calcium which was interpreted to mean increased calcium retention and new bone formation. However, the calcium appears to be deposited as calcium phosphate in both soft and hard tissue while the postprandial serum phosphate elevations cause secondary hyperparathyroidism (Jowsey et al., 1974).

Estrogens and calcium supplements will arrest bone resorption but do not stimulate bone formation and therefore will not result in the increased bone mass that is necessary for cessation of fractures, unless stress is decreased (Riggs et al., 1969). Fluoride alone appears to cause an increase in bone mass. This would appear to be beneficial since an effective treatment must not merely stop bone resorption or bone loss at the stage at which the first fracture occurs but must stimulate bone formation. Preventive treatment with either hormones or calcium supplements may be useful if initiated before the mass of bone has decreased to a level at which minor trauma will cause fracture. However, these agents will not increase bone mass, and any symptomatic improvement is probably the result of decreased stress.

It is necessary, perhaps, to point out the difference between a two to five year fluoride treatment of a bone disease, in an adult, and chronic, life-long ingestion of fluoride. Nevertheless, it was largely observations of osteosclerosis in areas of endemic fluorosis (Figure 1) that stimulated interest in the potential of fluoride in the treatment of osteoporosis (Jolly et al., 1968). Fluoride has been used in the past for the treatment of osteoporosis or bone losing diseases such as multiple myeloma and hypercortisonism (Cohen et al., 1969). The early studies suggested that the bone mass did not always increase; that fracture rate did not decrease, and even though the calcium balance was improved during the early part of the treatment period, a positive calcium balance was not maintained (Rich et al., 1964). There was, however, strong evidence from the bone biopsy

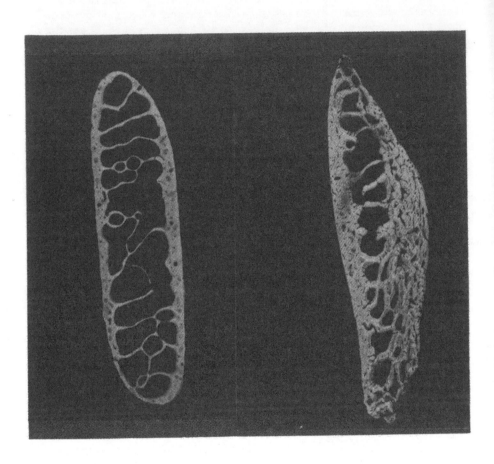

Figure 1. Microradiographs of cross-sections of ribs from (left) a normal 50-year-old male and (right) and elderly male from Punjab, India, an area of endemic fluorosis. Note the increased thickness of cortical bone and of the trabeculae, which has resulted from a life-long exposure to high levels of fluoride in the drinking water and food (magnification x6).

material that new bone was being deposited in greater amounts than normal. The biochemical findings were a decrease in serum calcium and an increase in alkaline phosphatase, both of which are indicative of osteomalacia (Jowsey et al., 1968). The bone biopsy specimens also showed signs of osteomalacia; the unmineralized osteoid which bordered the new bone and areas of poorly mineralized bone were evident in the micro-radiography and mineralized stained section. There were also areas of unmineralized bone around the osteocytes (Figure 2). Other more recent studies demonstrate the development of secondary hyperparathyroidism. Apparently fluoride stimulates osteoblasts to lay down new matrix which increases the amount of calcium needed to mineralize bone. The average American diet is deficient in calcium in that the daily intake is not sufficiently high to prevent a negative calcium balance (Heaney et al., 1974). The diet also contains excessive amounts of phosphorus; i.e., there is adequate phosphorus for new bone mineralization but not adequate calcium. As a result of the increased demand for calcium, serum calcium tends to decrease and the new bone remains unmineralized, giving the morphological appearance of osteomalacia. Also, as a result of the decrease in serum calcium, there is a stimulation of parathyroid hormone secretion and increased bone resorption. Fluoride alone will thus produce osteomalacia and increase bone resorption. These considerations led us to believe that if calcium were added to the fluoride regimen, a positive calcium balance might be reestablished allowing the increased volume of new bone to be properly mineralized and also preventing stimulation of parathyroid gland activity.

In our initial studies patients were treated with various levels of both fluoride and calcium. There was no significant decrease in serum calcium or elevation of alkaline phosphatase suggesting that osteomalacia had been avoided by the administration of calcium with the fluoride (Jowsey et al., 1972). There was a relationship between the levels of fluoride and calcium in these patients and the changes in bone resorption and formation rates that occurred during the treatment period. If the amount of supplemental calcium was 300 mg per day or less, the increase in resorption that occurred with fluoride alone was not prevented. Stimulation of resorption was prevented or decreased in patients given calcium supplements of 600 mg per day or more. The dose of sodium fluoride varied from 15 to 90 mg per day. Bone formation was increased in the majority of patients. There was a relationship between the change in bone formation and the level of fluoride which allowed us to predict that with a dose of about 25 mg per day or more of fluoride ion there would be significant increase in new bone formation.

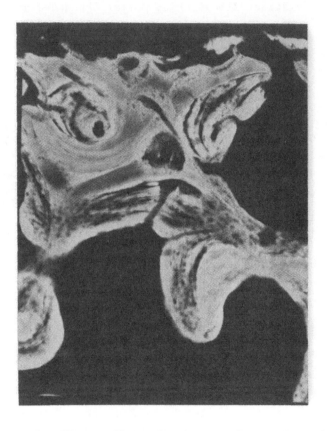

Figure 2. Microradiograph of a section of bone from the
iliac crest of a patient with osteoporosis treated with high
doses of fluoride, approximately 120 mg per day, and no
calcium or vitamin D. Areas of (pretreatment) normal bone
can be seen with new bone laid down on the surface which
characteristically shows both hypermineralized and hypomin-
eralized areas. The areas of low mineral density have
resulted from inadequate calcium intake (magnification x64).

No relationship was found between bone fluoride content and either serum fluoride values or bone surface formation changes; apparently the amount of fluoride in the bone reflects the life-time exposure of the particular patient to fluoride, and the period of treatment of one or three years had very little effect beyond slightly raising the very variable level of fluoride in the bone. X-rays of the lumbar and thoracic spine also reflected the increase in bone density. There was an increase in the thickness of the end plates of the vertebrae and also an increase in "trabeculation" caused by a thickening of spongy bone in the vertebral bodies. Similar findings in the lumbar spine have been reported by Hansson and Roos (1976) who treated osteoporotic patients for 18 months with fluoride and calcium and found a mean increase of 20% in vertebral density, as measured by a dual energy absorption method. Since both cortical and trabecular thickness are increased, the bone is not returned precisely to a normal situation because there is no increase in the number of new trabeculae (Figure 3). Also, the new bone tends to be hypermineralized, when formed during fluoride and calcium supplementation. Occasionally there are areas of poor mineralization around osteocytes (Figure 2).

One consideration in the treatment of osteoporosis with fluoride and calcium is the existence of side effects. With doses of 50 to 75 mg per day of sodium fluoride, fasting serum levels between 10 to 15 μM F are reached which are effective in increasing bone formation but may also produce unwanted reactions. We attempted to evaluate changes in the incidence of joint pain and swelling as well as gastrointestinal symptoms. Our main approach in the evaluation was to record symptoms as they occurred in the patients. Over a three year period, in eleven patients, four experienced mild arthralgia and joint stiffness, and six complained of epigastric discomfort; one patient suffered hair loss and one developed a bony spur over the bone biopsy site in the iliac crest. However, it is difficult to establish whether all these symptoms were the result of fluoride and calcium, or rather were symptoms that occur frequently in older women or in women with osteoporosis. For this reason, further studies into the nature and incidence of side effects are being carried out.

Studies of the therapeutic effectiveness of fluoride and calcium are continuing; the high doses of vitamin D that were used in the first study have not been used in subsequent studies. Rather, vitamin D deficiency is avoided by giving a dose equivalent to the minimal daily requirement since elderly people in Minnesota may well become somewhat vitamin D deficient during the winter months. The levels of sodium fluoride vary between 60 and 85 mg per day and 1.3 grams of

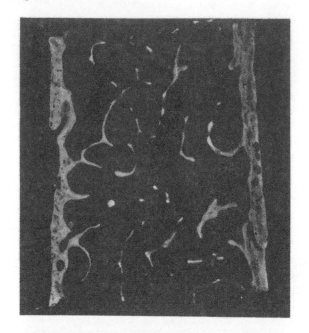

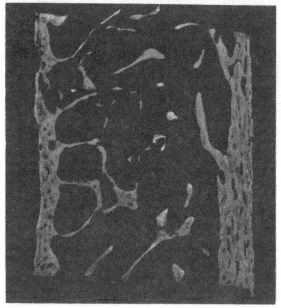

Figure 3. Microradiographs of the iliac crest of a patient before (top) and after (bottom) one year of treatment with fluoride and calcium (65 mg sodium fluoride and 3.0 grams calcium carbonate per day). There has been a modest increase in bone mass in both cortical and trabecular areas (magnification x8).

calcium per day is given, in the form of calcium carbonate. Fracture rate, side effects and changes in bone mass are also being studied in the current investigations.

Parallel with the studies in osteoporotic patients, animal studies are being carried out to evaluate the strength of bone in animals given fluoride and calcium. Studies in rats given fluoride and fed a diet with a ratio of calcium to phosphorus greater than one showed increase in bone mass and no decrease in the response of bone to stress (Guggenheim et al., 1976). In man large increases in bone mass resulting from chronic exposure to fluoride have been related to proportionally large increases in bone strength (Franke et al., 1976). Other experimental studies, largely in growing rats or birds, have used a high phosphorus diet to accompany fluoride and have not succeeded in increasing bone strength but rather have decreased it (Riggins et al., 1976). These studies emphasize the importance of accompanying the fluoride with amounts of calcium sufficient to prevent osteomalacia and secondary hyperparathyroidism.

Concluding Remarks

The preponderance of evidence suggests that administration of fluoride and calcium is an effective and relatively safe form of treatment for osteoporosis and would result in addition to the skeleton of a measurable amount of mechanically sound mineralized bone, thus reversing the osteoporotic process and preventing further fractures. Although some side effects may accompany fluoride therapy, they are relatively few and are probably preferable to the often disabling consequences of progressive osteoporosis.

This investigation was supported in part by Research Grant AM-8658 from the National Institutes of Health, Public Health Service.

References

Chalmers, J., Conacher, W. D. H., Gardner, D. L. and Scott, P. J. (1967) Osteomalacia--A common disease in elderly women. J. Bone Jt. Surg. 49-B; 403-423.

Cohen, P., Nichols, G. and Banks, H. H. (1969) Fluoride treatment of bone rarefaction in multiple myeloma and osteoporosis. Clin. Orthop. and Rel. Res. 64;221-249.

Franke, J., Runge, H., Grau, P., Fengler, F., Wanka, C. and Rempel, H. (1976) Physical properties of fluorosis bone. Acta Orthop. Scand. 47;20-27.

Guggenheim, K., Simkin, A. and Wolinsky, I. (1976) The effect of fluoride on bone of rats fed diets deficient in calcium or phosphorus. Calc. Tiss. Res. 22;9-17.

Hansson, T. and Roos, B. (1976) Effect of combined therapy with sodium fluoride, calcium and vitamin D on the lumbar spine in osteoporosis. Am. J. Roentgenol. Radium Ther. Nucl. Med. 126;1294-1296.

Heaney, R. P., Recker, R. R. and Saville, P. D. (1974) Calcium balance and calcium requirements in middle-aged women. Clinical Research 22;649A.

Jolly, S. S., Singh, B. M. Mathur, O. C. and Malhotra, K. C. (1968) Epidemiological, clinical and biochemical study of endemic dental and skeletal fluorosis in Punjab. Brit. Med. J. 4;427.

Jowsey, J., Schenk, R. K. and Reutter, F. W. (1968) Some results of the effect of fluoride on bone tissue in osteoporosis. J. Clin. Endocrin. & Metab. 28;869-874.

Jowsey, J., Riggs, B. L., Goldsmith, R. S., Kelly, P. J. and Arnaud, C. D. (1971) Effects of prolonged administration of porcine calcitonin in postmenopausal osteoporosis. J. Clin. Endocrin. & Metab. 33;752-758.

Jowsey, J., Riggs, B. L., Kelly, P. J., Hoffman, D. L. and Bordier, Ph. (1971a) The treatment of osteoporosis with disodium ethane-1-hydroxy-1, 1-diphosphonate. J. Lab. Clin. Med. 78;574-584.

Jowsey, J., Riggs, B. L., Kelly, P. J., and Hoffman, D. L. (1972) Effect of combined therapy with sodium fluoride, vitamin D and calcium in osteoporosis. Amer. J. Med. 53; 43-49.

Jowsey, J., Reiss, E. and Canterbury, J. M. (1974) Long-term effects of high phosphate intake on parathyroid hormone levels and bone metabolism. Acta Orthop. Scand. 45;801-808.

Melick, R. A., Martin, T. J. and Storey, E. (1973) Use of porcine calcitonin in osteoporosis. Aust. N. Z. J. Med. 3; 752-758.

Rich, C., Ensinck, J. and Ivanovich, P. (1964) The effects of sodium fluoride on calcium metabolism of subjects with metabolic bone diseases. J. Clin. Invest. 43;545-556.

Riggins, R. S., Rucker, R. C., Chan, M. M., Zeman, F. and Beljan, J. R. (1976) The effect of fluoride supplementation on the strength of osteopenic bone. Clin. Orthop. & Rel. Res. 114;352-357.

Riggs, B. L., Jowsey, J., Kelly, P. J., Jones, J. D. and Maher, F. T. (1969) Effect of sex hormones on bone in primary osteoporosis. J. Clin. Invest. 48;1065-1072.

5

Inorganic and Organic Fluorine in Human Blood

Warren S. Guy

> The quantitative determination of the amounts of
> fluorine contained in organic tissue is a matter
> of considerable difficulty. Anyone who has worked
> on the subject will endorse this statement. The
> difficulties lie both in the isolation and in the
> final determination of the element. A good many
> of the analytical methods employed in earlier
> works may be regarded as unreliable. . . . It is
> only within recent years that sensitive methods
> have been published combining simplicity with
> relative reliability.
>
> --Kaj Roholm, 1937

> Results from a new method of analysis of fluoride
> from serum indicate that the generally accepted
> value for normal humans is too high by as much as
> a factor of ten.
>
> --D. R. Taves, 1966

The above quotations underscore what has been a source
of frustration and controversy for scientists involved in
the quantitation of fluorine in biological samples for over
a century, namely, poor reproducibility, especially between
laboratories. Analytical results seem to vary with the
methods employed. The first two AAAS symposia on fluoride
(Moulton, ed., 1942, 1946) included no mention of this problem. In the last symposium in this series Largent (1954)
expressed only skepticism about the literature in this area:
"Analytical difficulties are frequently encountered in the
course of determining the fluoride content of biological
materials. This is true perhaps most frequently in the case
of blood . . . individual values for the concentration of
fluoride in blood must be viewed as somewhat undependable."

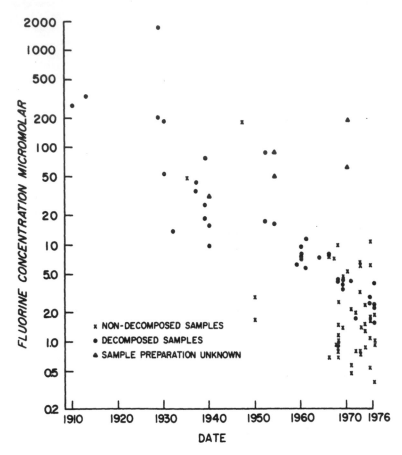

Figure 1.   Published Values for Fluorine Concentration in
Human Plasma Plotted Chronologically

Points represent the average of values reported. For
comparison's sake, values were excluded when reported as
zero, where subjects were reported to have drinking water
containing more than 2.5 ppm F or to be receiving more than
3.0 mg F per day from any source, for fasting subjects and
for subjects suffering from renal or skeletal disorders including fluorosis. Values reported for whole blood were
increased by a factor of 1.3, assuming 72% of the fluoride
in whole blood is in the plasma (according to the known
distribution of [18]F⁻, Carlson, et al., 1960a) and using 0.45
for the hematocrit. All values in this figure are referenced in the accompanying table except those for which
either no method was reported or the method referred could
not be located. Those not referenced in the table are
Wilkie (1940), Yudkin, et al. (1954) and Domzalska, et al.
(1970).

Within the past decade analytical methods have been greatly improved through two important developments: the observation that silicones facilitate separation of fluoride by diffusion (Taves, 1968e) and extraction (Fresen et al., 1968) and the introduction of the fluoride ion-specific electrode (Frant et al., 1966). These new methods have permitted the demonstration that human plasma contains both organic and inorganic forms of fluorine. They have also provided a foundation for a better understanding of the metabolism of inorganic fluoride (Taves and Guy, this monograph).

This manuscript will first compare published values for the concentration of fluoride in human blood and plasma on the basis of the analytical methods employed. The concentration of inorganic fluoride will be discussed in the light of information derived by this comparison, and methods for future work will be recommended. Information on organic fluorocompounds in human plasma will then be reviewed.

Published Values for Fluoride Concentration in Human Blood Compared by Analytical Method

Reported concentrations of fluoride in human blood have gradually decreased by three to four orders of magnitude over the past seven decades (Figure 1). In retrospect, it seems likely that improved specificity of methods and avoidance of selective contamination are responsible for the decline. These are discussed in turn below; then, with the perspective provided, the published values shown in Figure 1 are grouped and discussed according to analytical methods.

Problems of Specificity

Body fluids are complex, varying, and incompletely defined entities, making it difficult to test all potential interfering agents. Investigators, in testing their methods, commonly demonstrate good recovery of fluoride through any separation steps, an acceptable degree of sensitivity, and equivalent responses to small amounts of fluoride added to the sample. These tests, however, do not prove specificity. Negative interfering substances (making the value low) can still be present, although they would have to be present in amounts smaller than the amount of fluoride not to affect the recovery of stable fluoride added to the sample. Positive interfering substances, on the other hand, can still be present in any amount. The investigator may have eliminated certain substances known to cause positive interference, but others may exist and produce significant errors. The presence of interfering substances is suggested by

differences between the percent of the apparent fluoride and
the percent of radioactive fluoride recovered as a function
of time. It can also be inferred from discrepancies among
results obtained by different methods. These more sophisticated tests have been used only rarely. Even then the possibility of interference is not completely ruled out.

Problems Posed by Analytical Procedures

Nonspecificity. Fluoride in blood has been determined
by potentiometry (fluoride electrode), gas chromatography,
titration, colorimetry, and fluorimetry. These methods vary
in their specificity for fluoride. The fluoride electrode
has only one known positive interfering substance, hydroxide
ion (Butler, 1969), and the concentration of hydroxide ion
can be controlled by buffering. Gas chromatography can also
be considered to have a relatively high degree of specificity. The other three methods have a common basis in the
sense that each depends on the formation of a complex of
fluoride with a metal ion. Any substance in the analyzed
solution which also binds the metal ion gives positive
interference and leads to an exaggerated reading. The
increase is included in the blank only when interfering
agents arise from reagents or procedures which affect
samples and blanks equally. Interfering substances which
originate with the sample itself or which are preferentially
incorporated into the sample during the processing procedure
will appear as fluoride. Many components of human plasma
are known to be capable of interfering (Elving et al., 1954;
McKenna, 1951). Therefore, fluoride has to be separated
from the sample if reliable values are to be obtained. The
three most commonly used methods for fluoride separation
have been distillation, heat-facilitated diffusion, and
silicone-facilitated diffusion. Fluoride compounds are
volatilized from acid solution in each of these methods, but
at different temperatures, namely, 135°C, 60°C, and 25°C,
respectively. Volatile interfering agents are naturally
more likely to travel with fluoride-containing gases at
higher temperatures.

Contamination. In fluoride analysis selective contamination can occur through the use of glassware for collection, storage, and separation of serum (Singer and Armstrong,
1959; Taves, see Husdan, 1977). Theoretically, it can also
occur during decomposition of the samples in preparation for
fluoride analysis. Organic fluorocompounds such as freons
contaminating laboratory air may be decomposed on hot surfaces inside the furnace (F. A. Smith and D. R. Taves,
unpublished data). Samples and blanks may accumulate
fluoride to different degrees depending on their position

in the furnace and the surface area of the fluoride fixative
material in the crucibles. The surface area can be greatly
enhanced by the frothing that occurs when samples containing
protein are heated. Also, when no fixative salts are added,
samples accumulate more fluoride from furnace air than do
blanks because plasma contains calcium and magnesium salts
capable of fixing fluoride (W. S. Guy and D. R. Taves,
unpublished data).

Comparison of Published Methods and Results

With these general points about methods for fluoride
analysis in mind, consider the values published over the
years for the fluoride content of human blood. First, compare values for decomposed and non-decomposed samples
analyzed in the same ways. In all but one of twelve such
sample groups shown in Table 1, values for combusted samples
are markedly higher. (The exception is discussed below.)
Obviously, methods in which samples are prepared by combustion lack specificity for inorganic fluoride because
organic fluorine in the sample would be included. Total
fluoride has no biological meaning since there is no known
relation between organic and inorganic fluorine.

Values for fluoride levels in nondecomposed samples
fall into three ranges: low, 0.4-1.5 μM F⁻; medium, 2-11;
and high, 48-190. For the sake of comparison, values in the
table for fasting individuals and those known to have been
consuming excess fluoride (drinking water containing more
than 1.1 ppm F⁻) are excluded from this consideration.

Five methods produce results in the low range: 1) determination with the fluoride electrode, both when fluoride
is separated and concentrated by various means and, in most
cases, when fluoride is measured directly; 2) determination
by fluorimetry when fluoride is separated by silicone-
facilitated diffusion at 25°C; 3) determination by titration
with thorium nitrate when fluoride is separated by distillation, the distillate residue ashed, and fluoride redistilled;
4) determination by gas chromatography when fluoride is
separated by extraction as a fluorosilane; and 5) estimation
by calculation based on renal clearance of $^{18}F^-$ and normal
urine fluoride concentration. Three methods give values in
the medium range: 1) direct determination, in some cases,
with the fluoride electrode; 2) fluorimetry, and 3) colorimetry, both following diffusion at 60°C for 22 hours. Two
methods give values in the high range: titration with
thorium nitrate and colorimetry, both following separation
by distillation. The methods in each category will not be
examined more closely.

TABLE 1. PUBLISHED VALUES FOR THE CONCENTRATION OF FLUORINE IN HUMAN PLASMA[a]

Analytical method	Ashed samples μM	Ashed samples (n)	Nonashed samples μM	Nonashed samples (n)	Drinking water F (ppm)	Sample Population	Reference
A. Gravimetric, of precipitate	270	(1)	--	--	--	Adult male	Zdarek, 1910
B. Indirect colorimetric (PbS pre-cipitated from PbF_2 solution)							Steiger, 1908
	330	(1)	--	--	0.17	Accident victim	Gautier & Clausmann, 1913
C. Colorimetric (on ether extracted blood)	1800	(3)	--	--	0.03-0.17	Hemophiliacs	Stuber & Lang, 1928, 1929
	210	(15)	--	--	0.17	Other patients	
	0	(1)	--	--	0.03	Normal human	
	0	(1)	--	--	--	Hemophiliac	Hoff & May, 1930
	0	(1)	--	--	--	Normal human	
	54	(2)	--	--	--	Hemophiliacs	Feissly et al., 1930
	187	(3)	--	--	--	Normal humans	
	14	(1)}	--	--	--	Hemophiliacs	Brandes, 1932
	0	(2)}	--	--	--		
	0	(4)}	--	--	--	Normal humans	
	14	(1)}	--	--	--		
D. Titration of distillate							
1. Zirconium and purpurin; distillate ashed or twice redistilled	36	(4)	--	--	0.14	Basedow patients	Kolthoff & Stansby, 1934
	45	(3)	--	--	0.14	Other patients	
	80		--	--	--	Basedow patients	Kraft and May, 1937
	19		--	--	--	Normal humans	Wulle, 1939
	16	(10)	--	--	--	Normal humans	Hartmann et al., 1940
2. Thorium and sodium alizarin sulfonate	26	(4)	--	--	--	--	Gettler & Ellerbrook, 1939
	10	(1)	--	--	--	--	von Fellenberg, 1948
	18	(13)	--	--	0.10	Mothers at delivery	Held, 1952
	14	(16)	--	--	0.10	Mothers at delivery	Held, 1954

[a] Values (average) reported for serum are included unchanged; but, for the sake of comparison, values for whole blood are increased by a factor of 1.3 on the assumption that 72% of the fluoride in whole blood is in the plasma, according to the distribution of ^{18}F added to whole blood (Carlson et al., 1960a), and that the hematocrit value is 45. Values for subjects reported to suffer from renal or skeletal disorders or acute fluoride poisoning, or to be under fluoride therapy were excluded.

[b] Combustion (ashing with a fluoride fixative at 400-600°C except where noted) was the first step.

Table 1 (Continued)

Analytical method	Ashed samples		Nonashed samples		Drinking water F (ppm)	Sample population	Reference
	μM	(n)	μM	(n)			
D. (continued)							
3. Thorium and Zr alizarin	--		190	(20)	--	Normals, Scotland	Willard & Winter, 1933 Bowler et al., 1947
4. Thorium and sodium alizarin sulfonate; distillate ashed and redistilled	--		0 1.9	(30) } (4) }	0.06	Fasting adults	Smith and Gardner, 1951
	--		0 1.7	(9) } (11) }	0.06	Nonfasting adults	Smith et al., 1950
	--		0 2.9	(1) } (11) }	1.36	Nonfasting adults	
5. Thorium and chrome azurol S	93[d]	(1)	--		--	--	Waldo & Zipf, 1952
E. Colorimetric after gaseous separation							
1. Zirconium purpurin; distillation	--		48	(79)	--	Patients	Kolthoff and Stansby, 1934 Goldemberg & Schraiber, 1935
2. Zirconium eriochrome cyanine R; distillation	6.5	(2)	--		--	--	Singer and Armstrong, 1959
	7.3	(16)	--		0.15	Blood donors	
	7.8	(18)	--		1.10	Blood donors	Singer and Armstrong, 1960
	10.0	(36)	--		1.10	Blood donors	
	8.3	(26)	--		2.50	Blood donors	
	15.0	(22)	--		5.40	Blood donors	
	4.6	(18)	--		--	Adults	Singer and Armstrong, 1969
3. Violet iron salicylate complex; distillation	6.1	(25)	--		0.55	Mothers at term	Cremer & Voelker, 1953
	12.0	(37)	--		0.55	Nonpregnant females	Gedalia et al., 1961
	7.8	(39)	--		0.10	Mothers at term	Gedalia et al., 1964
4. Lanthanum alizarin, diffusion at 60°C for 24 h	8.3		--		--	--	Hall, 1963 Takaesu, 1966

d Combustion in a peroxide bomb.

Table 1 (Continued)

Analytical method	Ashed samples μM	(n)	Nonashed samples μM	(n)	Drinking water F (ppm)	Sample population	References
E. (continued)							
5. Zr eriochrome cyanine R; diffusion at 60°C for 22 h	3.6	(18)	7.3	(1)	--	--	Singer & Armstrong, 1959,1965
	--		4.9	(18)	--	Adults	Singer & Armstrong, 1967
	--		5.5	(27)	--	Mothers at delivery	Singer & Armstrong, 1969
	--		11.0	(2)	--	--	Armstrong & Singer, 1970
							Venkateswarlu, 1975b
6. Lanthanum alizarin; ion-exchange	--		2.6	(10)	1.0	2.0 mg F/day	Cox & Dirks, 1968
	--		3.8	(10)	1.0	4.7 mg F/day	
7. Zr eriochrome cyanine R; diffusion at 60°C for 22 h, ultrafiltrate	--		7.8	(18)	0.1	Adults	Singer & Armstrong, 1969,1965
							Singer & Armstrong, 1973
8. Cerium alizarin; diffusion at 60°C for 22 h	--		6.2	(2)	--		Baumler & Glinz, 1964
							Venkateswarlu, 1975b
F. Fluorimetric with morin thorium after gaseous separation							
1. Diffusion at 60°C for 22 h	--		6-10	(6)	1.0	Adults	Singer et al., 1965
	--		10	(8)	1.0	--	Taves, 1966
							Taves, 1968a
2. Silicone-facilitated diffusion at room temperature	--		0.7	(16)	1.0	Adults	Taves, 1968d,e
	4.4	(4)	0.7	(5)	1.0	--	Taves, 1966
	--		0.85	(3)	1.0	--	Taves, 1968a
	4.4	(1)	2.2	(1)	1.0	Patient	Taves, 1968b
	--		0.88	(16)	1.0	Mothers at delivery	Taves, 1971
							Shen and Taves, 1974
G. Fluoride estimated on basis of normal renal clearance of ^{18}F of 50 ml/min and normal urine F concentration of 50 μM	--		0.9-1.0		1.0	Normal adult	Taves, 1967, 1968b

Table 1 (Continued)

Analytical method	Ashed samples μM	Ashed samples (n)	Nonashed samples μM	Nonashed samples (n)	Drinking water F (ppm)	Sample population	Reference
H. Gas chromatography after extraction as fluorosilane	1.5	--	1.2	(5)	--	--	Fresen et al., 1968
			0.4	(9)	--	Normals	Belisle & Hagen, 1978
I. Electrode; no separation							
1. Buffered serum ultrafiltrate or plasma supernatant; standards in buffered saline		--	1.0	(12)	0.1	--	Barnes & Runcie, 1968
		--	0.78	(18)	0.1	Adults	Singer & Armstrong, 1973
		--	1.5	(2)	--	--	Venkateswarlu, 1974
		--	1.8	(5)	--	--	Venkateswarlu, 1975c
2. Fluoride added to serum defluorinated by adsorption to Ca3(PO4)2 to duplicate reading of original sample		--	0.57	(2)	--	--	Venkateswarlu et al., 1971
3. Single-known-addition		--	2.0	(26)	1.0	Preadm. surg. pts.	Hall et al., 1972
		--	0.54	(20)	0.18	Normals	Fuchs et al., 1975
4. Serum and standards diluted 1:1 with TISAB		--	3.3	(14)	0.15	No fluorosis	Jardillier & Desmet, 1973
		--	6.7	(28)	3.8	Dental fluorosis	
		--	2.4	(41)	1.0	In-patients, fasting	Parkins et al., 1974
		--	1.0	(136)	1.0	Normals, various ages	Husdin et al., 1976
5. Serum diluted 1:17 with acetate buffer		--	0.8	(25)	[non-fl.]	Patients	Fry & Taves, 1970
		--	1.4	(1937)	1.0	Patients	Inkovaara et al., 1973
		--	1.3	(1083)	1.0	Fasting patients	Hanhijärvi, 1974
		--	0.88	(501)	0.2	Fasting patients	
		--	1.5	(122)	0.1	Fasting	Kuo & Stamm, 1975
J. Electrode; fluoride separated and concentrated into small volume							
1. Diffusion at 60°C for 22 h	4.0	(18)	1.5	(8)	1.0	--	Singer & Armstrong, 1965
			1.4	(18)	--	Adults	Taves, 1968a
			1.7	(2)	--	--	Singer & Armstrong, 1965
		--					Venkateswarlu, 1975b

Table 1 (Continued)

Analytical method	Ashed samples μM	(n)	Nonashed samples μM	(n)	Drinking water F (ppm)	Sample population	Reference
J. (Continued)							
2. Silicone-facilitated diffusion at room temperature	4.5	(3)	0.9	(1)	1.0	--	Taves, 1968d, 1968e
	--		0.88	(1)	1.0	--	Taves, 1968a
	1.8	(65)	0.8	(65)	1.0	Blood donors	Taves, 1968b
	1.6	(30)	0.38	(30)	0.1	Blood donors	Guy, 1972
	2.3	(12)	1.0	(12)	0.9	Blood donors	Guy et al., 1976
	2.5	(30)	0.89	(30)	1.0	Blood donors	
	4.2	(4)	1.9	(4)	2.1	Blood donors	
	5.4	(30)	4.3	(30)	5.6	Blood donors	
3. Adsorption onto Ca$_3$(PO$_4$)$_2$	--		0.47	(4)	--	--	Venkateswarlu et al., 1971
	--		1.3	(1)	--	--	Venkateswarlu, 1974, 1975a,b
4. Extraction as a fluorosilane	--		1.3	(1)	--	--	Venkateswarlu, 1974
	2.6	(2)	1.1	(2)	--	--	Venkateswarlu, 1975a,b
	3.0^e	(2)	--				

^{e}Combustion in an oxygen bomb.

Low range. Many of the values in the low range were
determined with the fluoride electrode (Sections I and J in
Table 1). In each case interference by hydroxide was controlled by buffering the reading solution. When fluoride
was first separated from the sample and concentrated into a
small volume, recovery checked with $^{18}F^-$ was reported to be
high (97-98%) (Taves, 1968b).

The second method yielding low-range values is fluorimetric determination with morin thorium when fluoride was
separated by silicone-facilitated diffusion, a process performed at room temperature (Table 1, Section F2). Recovery
was reported to be high for both $^{18}F^-$ (98%) and nonradio-
active fluoride (97%). As evidence for specificity against
volatile interfering substances, values obtained after
separation by silicone-facilitated diffusion are reported
to be unaffected by diffusion times of 20-158 hours.

The third method is double distillation and titration
of fluoride with thorium nitrate (Section D4, Table 1).
High-range values were obtained when fluoride was separated
by a single distillation only (Section D1-3). A certain
degree of skepticism is warranted since, in contrast to the
other methods yielding low-range results, blanks were several times greater than the net amounts of fluoride being
measured, a large percentage of values obtained were reported as zero, and recovery of added fluoride was only
about 80%.

In the fourth method yielding low-range values, fluoride was determined by flame ionization following separation
by extraction as a fluorosilane and gas chromatography (Section H). These two separation steps are probably very efficient, and recovery of added fluoride was high (>92%).

Thus, of these four independent analytical methods
yielding low-range values for fluoride in human plasma, only
one is questionable. All would be expected to have a relatively high degree of specificity for fluoride. Values in
the low range are corroborated by calculations (Section G,
Table 1) based on the measured renal clearance of $^{18}F^-$ and
on undisputed values for the average concentration of fluoride in urine. These low-range values have been obtained in
nine independent laboratories.

Medium and high ranges. Values in the medium and high
ranges were obtained with relatively less specific or otherwise questionable methods. Direct measurement of fluoride
in blood of normal individuals with the fluoride electrode
(Section I, Table 1) is difficult because the fluoride

activity is below the limit of linearity of the electrode response, generally accepted to be about 1 μM F activity. (In plasma, fluoride activity is significantly lower than fluoride concentration because of the high ionic strength and the presence of certain fluoride-binding agents such as aluminum, calcium, and magnesium ions.) Fluoride activity is further reduced by dilution of the samples with buffering solutions. Where the electrode response is below its linear range, the apparent fluoride concentration obtained from any given voltage reading on an assumed straight-line response curve will be erroneously high.

This problem of converting electrode response to fluoride concentration when the relationship is not linear may explain the medium-range values obtained by Hall et al. (1972). However, their samples were not from normal subjects and may be high for that reason. Fuch et al. (1975), using a similar technique, compensated for the nonlinearity problem by what they called the "electrode slope-by-dilution method" and obtained low-range values. Another problem is one of preparing appropriate standards and blanks. Jardillier et al. (1973), for example, made no attempt to duplicate the ionic character of serum in the standard solutions. Also, at low fluoride activity levels the electrode response is sluggish and suffers from a "memory" of previously applied solutions. Some investigators doing direct analysis took their readings at arbitrarily set times and did not explain how they avoided memory problems.

Interfering substances for colorimetric and fluorimetric reagents, probably of an organic nature, accumulate in the trapping solution during fluoride diffusion at 60°C. Average values obtained in this way vary from 4.9 to 11 μM F⁻ (Sections E5,7,8 and F1, Table 1). Where the ion-specific electrode is used for determination of fluoride in the same samples the average values reported vary from 1.4 to 1.7 (section J1). Fusing the diffusion receiver contents with Na_2CO_3 dropped the amount apparent by fluorimetry to one-tenth its original value despite retention of 90% of [18]F⁻ added as a control (Taves, 1966).

Some authors have chosen to designate values obtained by diffusion-colorimetry method "total fluoride" because of their agreement with values for ashed samples. They emphasize that the diffusion time must be exactly 22 hours (Singer et al., 1969, 1973). At diffusion times longer than 22 hours higher values are obtained and at shorter diffusion times lower values are obtained (Singer et al., 1967, 1969). Diffusion of [18]F⁻ by this method, however, is more than two-thirds complete in 2 hours (Taves, 1968d). Venkateswarlu

(1975b) in his elaborate and convincing criticism of both method and terminology writes, "The values so obtained reflect ionic fluoride plus interfering substances, the latter masquerading as nonionic fluorine. . . . The continued use of the diffusion colorimetric procedures for the determination of total F and nonionic F in unashed body fluids would yield more misleading information; and this practice should be discontinued."

The extremely high values obtained by titration with thorium nitrate (Section D3, Table 1) and colorimetry (Section E1) of blood distillates probably reflect an extension of the problem with diffusion at 60°C. Any interfering substance capable of codiffusing with fluoride at 60°C could certainly codistill at 135°C in even greater amounts.

Medium- and high-range values, obtained by colorimetry, fluorimetry, and titration methods where fluoride was separated from the sample by distillation or heat-facilitated diffusion, could all be too high because of a single compound which becomes volatile in acid solution and which acts as a positive interfering agent in the reading solutions.

Validity of Results

No single type of error could logically have lowered all values obtained by the five low-range methods. Except as discussed above, these methods were shown to have high recovery of added fluoride and adequate sensitivity. A negative interference, if postulated, would have to affect in the same way measurement systems as diverse as the fluoride electrode, the flame ionization detector, and fluorimetric and titrimetric reagents.

The possibility is remote that all the published values for fluoride in human blood and plasma are correct and reflect true differences in the samples analyzed. Most of the high published values came from Germany and other European countries through the 1940s, and thereafter most came from the United States; but there is no obvious reason why these geographic and temporal differences should so greatly affect the fluoride concentrations of the blood of normal individuals as to cause differences of several orders of magnitude. Some investigators have compared values obtained on identical samples by different methods, and, in general, values obtained from low-range methods (fluorimetry and fluoride electrode) agreed, whereas those from medium-range methods were higher (Taves, 1968b; Singer et al., 1969;

Fry et al., 1970; Venkateswarlu, 1975b; W. S. Guy and D. R. Taves, unpublished).

In conclusion, of all the published values for the fluoride concentration in human plasma, those in the low range, 0.4-1.5 μM, for nondecomposed samples are probably nearest the true mean value for normal individuals, since these values were obtained with several independent and relatively fluoride-specific methods. Recommendations on methods for future work would include the following: a) avoid sample decomposition, b) separate fluoride from the sample and concentrate it into a smaller volume by diffusion or extraction, and c) determine fluoride with the ion-selective electrode. Other methods should at least be tested against the above approach before their routine use.

Organic Fluorocompounds in Human Plasma

Identification. Evidence for the existence of organic compounds of fluorine in human plasma has been provided by Guy, Taves, and Brey (1976). They isolated an organic fluorine-containing fraction from plasma and characterized it by fluorine nuclear magnetic resonance spectroscopy. The isolation was accomplished by first dialyzing 20 liters of pooled plasma, freeze-drying the residual material and extracting it with methanol, fractionating the extract by liquid-liquid partition chromatography on a Sephadex column, and then separating components of the polar lipid fraction by silicic acid chromatography. The fluorine-containing part of the compounds in the major peak of the final chromatograms resembled perfluoro-octanoic acid.

Derivatives of perfluoro-octanoic acid are widely used as surfactants (Bryce, 1964), but no toxicological studies of these compounds appear to have been published, and a list of current products in which these compounds are incorporated is not generally available. These compounds have been generally introduced since 1940. Therefore, samples analyzed earlier would probably not contain much organic fluorine.

The chemistry of derivatives of perfluoro-octanoic acid (Bryce, 1964) is in keeping with the known characteristics of organic fluorine from human blood plasma. The solubility is like that of a very polar lipid. This is consistent with extraction by methanol and also with poor recovery in attempts to extract organic fluorine directly from plasma with solvents of low polarity such as heptane, petroleum ether, and ethyl ether (Guy, 1972). Compounds of this nature would be expected to bind to albumin because of their similarity to the fatty acids which are normally bound.

Such binding is consistent with the retention of organic
fluorine during dialysis and ultrafiltration and the migration of organic fluorine with albumin during electrophoresis
(Taves, 1968a,c; Guy, 1972; Venkateswarlu, 1975a; Guy et al.
1976).

Human plasma may contain other types of organic fluorocompounds. Guy et al. (1976) found multiple fluorine-
containing peaks in silicic acid chromatograms from the
major Sephadex fraction. The major peak contained 36% of
the original organic fluorine, and up to four other peaks
contained a total of 13%. The smaller peaks varied in size
relative to the major peak. About 20% of the original
organic fluorine appeared in three other Sephadex fractions.
It has not been determined whether or not these accessory
peaks and components of other fractions actually represent
different types of organic fluorocompounds.

Geographical distribution. The presence of small
amounts of organic fluorine appears to be widespread in
human plasma. The consistency with which values for fluoride in decomposed samples are found to be higher than corresponding values in noncombusted samples is evidence for
this. Values in Table 1 (not including those in Section E5)
represent averages for 167 samples taken from individuals
living in six U.S. cities in three states. In one study
(Guy et al., 1976; Section J of Table 1) samples from 106
individuals living in 5 cities in 2 states were analyzed.
A difference in fluoride concentration between ashed and
unashed samples was measurable for 104 individuals. The
two not measurable were samples having relatively high
inorganic fluoride concentrations (over 7 µM), making any
difference difficult to measure.

Organic fluorine content of blood appears to be independent of the fluoride concentration of the water supply.
The 5 cities mentioned above had water with fluoride concentrations varying from 0.1 to 5.6 ppm. The average
differences between ashed and unashed samples for each city
did not vary according to the fluoride content of the water.

In normal individuals drinking fluoridated or low-
fluoride water, organic fluorine constitutes the major
portion of total plasma fluorine. Values in Table 1 for
combusted samples are in most cases at least double the
values for corresponding noncombusted samples. These estimates are on the low side, because not all the organic
fluorine is recovered as inorganic fluoride where samples
are prepared by open ashing. For example, only 21% of the
fluorine atoms of perfluoro-octanoic acid added to plasma
was fixed as inorganic fluoride with the ashing procedure

used by Guy et al. (1976). Combustion in a closed compartment, such as a bomb or pyrohydrolysis assembly, leads to
more complete decomposition of organic fluorocompounds and
trapping of the generated inorganic fluoride. Kakabadse
et al. (1971) found that ashing of perfluoro-octanoic acid
led to 31% recovery of fluorine as inorganic fluoride while
pyrohydrolysis resulted in 100% recovery. Belisle et al.
(1977) obtained 92% recovery of fluorine after combusting
the same component in an oxygen bomb. Thus, data for
organic fluorine obtained with techniques involving open
ashing must be taken as minimum values.

Venkateswarlu (1975a) compared the amount of organic
fluorine apparent in the same serum samples combusted
either in the oxygen bomb or by open ashing. He found that
combustion in the bomb made 1.5 and 1.05 times more fluoride
apparent by ion-specific electrode analysis of two samples
from humans. He reported 2.2 and 1.4 μM organic fluorine
for the sample combusted in the bomb. Also with the bomb he
found no organic fluorine loss when samples were lyophil-
ized, showing that volatile organic compounds of fluorine
were not present in measurable quantities. Belisle et al.
(1977) found an average of 1.16 ± 0.35 μM organic fluorine
in 9 plasma samples analyzed by an oxygen bomb—gas chro-
matograph method. They generally found higher organic
fluorine concentrations in whole blood (1.72 ± 0.77 μM) than
in plasma.

Presence in lower species. The question of whether or
not natural organic fluorine compounds exist in the plasma
of lower species of animals has received only very limited
attention to date. Taves (1971) found less than 0.3 μM
organic fluorine in the plasma of 12 species of animals.
Guy et al. (1976) found 0.13 μM organic fluorine in one
dialyzed bovine plasma sample by ashing. However, on
silicic acid chromatography, no peaks corresponding to the
major peaks from human samples were observed for bovine
plasma. Venkateswarlu (1975a) found an average of 1.8 μM
organic fluorine in 5 bovine serum samples with the oxygen
bomb. The higher values obtained with the oxygen bomb
technique make this the combustion method of choice for any
further systematic search for organic fluorine. In respect
to the possible occurrence of organic fluorine in human
tissues, it should be mentioned that certain fluoro-organic
acids have been claimed to occur in forage crops exposed to
atmospheric fluoride pollution. Cheng et al. (1968) found
fluoroacetate and fluorocitrate in soybean leaves. Lovelace
et al. (1968) and Ming-Ho et al. (1970) found the same compounds in crested wheatgrass. However, Weinstein et al.
(1972) were unable to repeat these results. Ther:lore, the

significance of these findings for man is not presently known although both compounds are generally recognized as being highly toxic.

In conclusion, human plasma does contain organic fluoro-compounds. The major type is probably a derivative of perfluoro-octanoic acid, presumably a synthetic environmental contaminant. The exact amounts of organic fluorine in human plasma are not known, but they are usually high compared to the amounts of inorganic fluoride. More work is needed to determine if there are other types of organic fluorocompound contaminants and to determine whether or not organic fluoride occurs naturally in animal tissues. In any event, until such time as a conversion of the organic fluorine to inorganic fluoride or vice versa is shown to occur, measurement of total fluoride by itself will have no biological interest.

Bibliography

Angmar-Mänsson, B., Ericsson, Y. and Ekberg, O. (1976) Plasma fluoride and enamel fluorosis. Calcif. Tiss. Res., 22; 77-84.

Armstrong, W.D. and Singer, L. (1970) Distribution of fluoride in body fluids and soft tissues. In Fluorides in Human Health, WHO, Geneva, pp. 94-99.

Armstrong, W.D., Singer, L. and Makowski, E.L. (1970) Placental transfer of fluoride and calcium. Am. J. Obstet. Gynecol., 107; 432-434.

Backer-Dirks, O.B. et al. (1974) Total and free ionic fluoride in human and cow's milk as determined by gas-liquid chromatography and the fluoride electrode. Caries Res., 8; 181-186.

Barnes, F.W. and Runcie, J. (1968) Potentiometric method for the determination of inorganic fluoride in biological material. J. Clin. Pathol., 21; 668-676.

Bäumler, J. and Glinz, E. (1964) Determination of fluoride ion in microgram amounts. Mitt. Lebensmitt. Hyg., 55; 250-264.

Belisle, J. and Hagen, D.F. (1978) Method for total fluorine content of whole blood, serum/plasma and other biological samples. In press.

Bowler, R.G. et al. (1947) The risk of fluorosis in magnesium foundries. Br. J. Ind. Med., 4; 216-222.

Brandes, W. (1932) Concerning the supposed presence of fluorine in the blood of hemophiliacs. Z. Klin. Med., 119; 504.

Bryce, H.G. (1964) Industrial and utilitarian aspects of fluorine chemistry. In Fluorine Chemistry, V; 297 (J.H. Simon, ed.), Academic Press, New York.

Butler, J.N. (1969) Thermodynamic studies. In Ion-Selective Electrodes (R.A. Durst, ed.), p. 148, NBS, Washington,D.C.

Cheng, J.Y. et al. (1968) Fluoroorganic acids in soybean leaves exposed to fluoride. Environ. Sci. Technol., 2; 367.

Cox, F.H. and Backer-Dirks, O.B. (1968) The determination of fluoride in blood serum. Caries Res., 2; 69-78.

Cremer, H.D. and Voelker, W. (1953) The determination of fluorine in bones and teeth. Biochem. Z., 324; 89-92.

Elving, P.J., Horton, C.A. and Willard, H.H. (1954) Analytical chemistry of fluorine and fluorine-containing compounds. In Fluorine Chemistry, II; 102-157 (J.H. Simons, ed.). Academic Press, New York.

Feissly, R., Fried, I.W. and Oehrli, H.A. (1930) Hemophilia and blood fluorine. Klin. Wochenschr., 10; 829-830.

Frant, M.S. and Ross, J.W. (1966) Electrode for sensing fluoride ion activity in solutions. Science, 154; 1553.

Fresen, J.A., Cox, F.H. and Witter, M.J. (1968) The determination of fluoride in biological materials by means of gas chromatography. Pharm. Weekblad, 103; 909-914.

Fry, B.W. and Taves, D.R. (1970) Serum fluoride analysis with the fluoride electrode. J. Lab. Clin. Med., 75; 1020-1025.

Fuchs, C. et al. (1975) Fluoride determination by ion selective electrodes: A simplified method for the clinical laboratory. Clin. Chim. Acta, 60; 157-167.

Gautier, A. and Clausmann, P. (1913) Fluorine in animal organisms. C.R. Acad. Sci., 157; 94, method 154; 1469, 154; 1670-1677.

Gedalia, I., Rozenzweig, K.A. and Sadeh, A. (1961) Fluorine content of superficial enamel layer and its correlation with the fluorine content of saliva, tooth age, and DMFT count. J. Dent. Res., 40; 865-869.

Gedalia, I. et al. (1964) Placental transfer of fluoride in the human fetus at low and high F-intake. J. Dent. Res., 43; 669-671.

Gettler, A.D. and Ellerbrook, L. (1939) Toxicology of fluorides. Am. J. Med. Sci., 197; 625-638.

Goldemberg, L. and Schraiber, J. (1935) The fluorine content of the fluids of the human organism in various pathological conditions. Rev. Soc. Argentina Biol., 11; 43.

Guy, W.S. (1972) Fluorocompounds of human plasma: Analysis, prevalence, purification and characterization. Doctoral Thesis, Univ. of Rochester, Rochester, N.Y.

Guy, W.S., Taves, D.R. and Brey,Jr., W.S. (1976) Organic fluorocompounds in human plasma: Prevalence and characterization. In Biochemistry Involving Carbon-Fluorine Bond (Robert Filler, ed.) Am. Chem. Soc. Symp. Ser., No. 28, pp. 117-134.

Hall, L.L. et al. (1972) Direct potentiometric determination of total ionic fluoride in biological fluids. Clin. Chem. 18; 1455-1458.

Hall, R.J. (1963) The spectrophotometric determination of submicrogram amounts of fluorine in biological specimens. Analyst, 88; 76-83.

Hanhijärvi, H. (1974) Comparison of free ionized fluoride concentrations of plasma and renal clearance in patients of artificially fluoridated and non-fluoridated drinking water areas. Proc.Finn.Dent.Soc., Suppl.3, 1-67.

Hartmann, H., Chytrek, E. and Ammon, R. (1940) Concerning the fluorine content of human blood. Hoppe Seyl. Z. Phys. Chem., 256; 52-58.

Held, H.R. (1952) The penetration of fluorine through the placenta and its presence in milk. Schweiz. Med. Wochenschr., 82; 297-301.

Held, H.R. (1954) Fluorine medication and blood fluorine. Schweiz. Med. Wochenschr., 84; 251-254.

Hoff, F. and May, F. (1930) On the problem of hemophilia and blood fluorine. Z. Klin. Med., 112; 558-567.

Husdan, H. et al. (1976) Serum ionic fluoride: Normal range and relationship to age and sex. Clin. Chem., 22; 1884-1888.

Inkovaara, J. et al. (1973) Fluoride and osteoporosis. Br. Med. J., 1; 613.

Jardillier, J.C. and Desmet, G. (1973) Study of serum fluorine and its combinations by a technique utilizing a specific electrode. Clin. Chim. Acta, 47; 357-363.

Jolly, S.S., Singh, B.M. and Mathur, O.C. (1969) Endemic fluorosis in Punjab. Am. J. Med., 47; 553-563.

Kakabadse, G.J. et al. (1971) Decomposition and the determination of fluorine in biological materials. Nature, 229; 626-627.

Kolthoff, I.M. and Stansby, M.E. (1934) Detection and estimation of small amounts of fluorine. Application of the zirconium purpurin test. Ind. Eng. Chem. Anal. Ed., 6; 118-121.

Konikoff, B.S. (1974) The bioavailability of fluoride in milk. J. La. Dent. Assoc., Fall, pp. 7-12.

Kraft, K. and May, R. (1937) Contribution to the biochemistry of fluorine. II. Determination in blood and water. Hoppe Seyl. Z. Phys. Chem., 246; 233-243.

Kuo, H.C. and Stamm, J.W. (1975) The relationship of creatinine clearance to serum fluoride concentration and the urinary fluoride excretion in man. Arch. Oral Biol., 20; 235-238.

Largent, E.J. (1954) Metabolism of inorganic fluorides. In Fluoridation as a Public Health Measure (J.H. Shaw, ed.), pp. 54-55, AAAS, Washington, D.C.

Lovelace, J., Miller, G.W. and Welkie, G.W. (1968) The accumulation of fluoroacetate and fluorocitrate in forage crops collected near a phosphate plant. Atmos. Environ., 2; 187-189.

McKenna, F.E. (1951) Methods of fluorine and fluoride analysis, I, II, III. Nucleonics, 8; 24-33, 9; 40-49, 9; 51-58.

Ming-Ho, Y. and Miller, G.W. (1970) Gas chromatographic identification of fluoroorganic acids. Environ. Sci. Tech., 4; 492-495.

Moulton, F.R., ed. (1942) Fluorine and Dental Health. AAAS, Washington, D.C.

Moulton, F.R., ed. (1946) Dental Caries and Fluorine. AAAS, Washington, D.C.

Nickles, M.J. (1856) The presence of fluorine in blood. C.R. Acad. Sci., 43; 885.

Parkins, F.M. et al. (1974) Relationships of human plasma fluoride and bone fluoride to age. Calcif. Tiss. Res., 16; 335-338.

Roholm, K. (1937) Fluorine Intoxication. Nyt Nordisk Forlag, Copenhagen, p. 257.

Shen, Y.W. and Taves, D.R. (1974) Fluoride concentration in the human placenta and maternal and cord blood. Am. J. Obstet. Gynecol., 119; 205-207.

Singer, L. and Armstrong, W.D. (1959) Determination of fluoride in blood serum. Anal. Chem., 31; 105-109.

Singer, L. and Armstrong, W.D. (1960) Regulation of human plasma fluoride concentration. J. Appl. Physiol., 15; 508-510.

Singer, L. and Armstrong, W.D. (1965) Determination of fluoride: Procedure based upon diffusion of hydrogen fluoride. Anal. Biochem., 10; 495-500.

Singer, L. and Armstrong, W.D. (1967) Normal human serum fluoride concentrations. Nature, 214; 1161-1162.

Singer, L. and Armstrong, W.D. (1969) Total fluoride content of human serum. Arch. Oral Biol., 14; 1343-1347.

Singer, L. and Armstrong, W.D. (1973) Determination of fluoride in ultrafiltrates of sera. Biochem. Med., 8; 415-422.

Smith, F.A., Gardner, D.E. and Hodge, H.C. (1950) Investigations on the metabolism of fluoride. II. Fluoride content of blood and urine as a function of the fluorine in drinking water. J. Dent. Res., 29; 596-600.

Smith, F.A. and Gardner, D.E. (1951) Investigations on the metabolism of fluoride. I. The determination of fluoride in blood. J. Dent. Res., 30; 182-188.

Smith, F.A. et al. (1960) The effects of the absorption of fluoride. AMA Arch. Ind. Health, 21; 330.

Steiger, G. (1908) The estimation of small amounts of fluorine. J. Am. Chem. Soc., 30; 219-225.

Stuber, B. and Lang, K. (1928) Concerning the basis of hemophilia. Z. Klin. Med., 108; 423-444.

Stuber, B. and Lang, K. (1929) Investigations on the theory of blood clotting. Biochem. Z., 212; 96-101.

Takaesu, I. (1966) Micro- and ultramicrodetermination of fluorine using lanthanum-alizarin complexan chelate and a modified microdiffusion bottle. Koku Eisei Gakkai Zasshi, 16; 1-23.

Taves, D.R. (1966) Normal human serum fluoride concentrations. Nature, 211; 192-193.

Taves, D.R. (1967) Use of urine to serum fluoride concentration ratios to confirm serum fluoride analysis. Nature, 215; 1380.

Taves, D.R. (1968a) Evidence that there are two forms of fluoride in human serum. Nature, 217; 1050-1051.

Taves, D.R. (1968b) Determination of submicromolar concentrations of fluoride in biological samples. Talanta, 15; 1015-1023.

Taves, D.R. (1968c) Electrophoretic mobility of serum fluoride. Nature, 220; 582-583.

Taves, D.R. (1968d) Effect of silicone grease on diffusion of fluoride. Anal. Chem., 40; 204-206.

Taves, D.R. (1968e) Separation of fluoride by rapid diffusion using hexamethyldisiloxane. Talanta, 15; 969-974.

Taves, D.R. (1971) Comparison of "organic" fluoride in human and nonhuman serums. J. Dent. Res., 50; 783.

Venkateswarlu, P., Singer, L. and Armstrong, W.D. (1971) Determination of ionic (plus ionizable) fluoride in biological fluids: Procedure based on adsorption of fluoride ion on calcium phosphate. Anal. Biochem., 42; 350-359.

Venkateswarlu, P. (1974) Reverse extraction technique for the determination of fluoride in biological materials. Anal. Chem., 46; 878-882.

Venkateswarlu, P. (1975a) Determination of total fluorine in serum and other biological materials by oxygen bomb and reverse extraction techniques. Anal. Biochem., 68; 512-521.

Venkateswarlu, P. (1975b) Fallacies in the determination of total fluorine and nonionic fluorine in the diffusates of unashed sera and ultrafiltrates. Biochem. Med., 14; 368-377.

Venkateswarlu, P. (1975c) A micro method for direct determination of ionic fluoride in body fluids with the hanging drop fluoride electrode. Clin. Chim. Acta, 59; 277-282.

von Fellenberg, T. (1948) On the problem of the significance of fluorine for the teeth. Mitt. Gebiete Lebensmitt. Hyg., 39; 124-182.

Waldo, A.L. and Zipf, R.E. (1952) The determination of fluoride in drinking water and biological materials. J. Lab. Clin. Med., 40; 601-609.

Weinstein, L.H. et al. (1972) Studies on fluoro-organic compounds in plants. III. Comparison of the biosynthesis of fluoro-organic acids in Acacia georginae with other species. Environ. Res., 5; 393-408.

Willard, H.H. and Winter, O.B. (1933) Volumetric method for determination of fluorine. Ind. Eng. Chem., Anal. Ed., 5; 7-10.

Wulle, H. (1939) Contribution to the microdetermination of blood fluorine. Hoppe Seyl. Z. Phys. Chem., 260; 169-174.

Zdarek, E. (1910) Concerning the distribution of fluorine in particular human organs. Hoppe Seyl. Z. Phys. Chem., 69; 127-137.

6

Is Fluoride Intake in the United States Changing?

Donald R. Taves

Several published studies suggest that the average total intake of fluoride has increased in recent years. Spencer's group estimated an intake of at least 3 mg/day rather than the earlier estimates of 1.5 mg/day, primarily because of higher levels of fluoride in food (Osis et al., 1974). Balance data suggest a higher retention by bone, nearly 2 mg/day (Spencer et al., 1970) rather than 0.2 mg as expected from the amount accumulated in bone from a lifetime in a community with fluoridated water, [(2000 mg/kg bone ash)x(2.6 kg bone ash)/(70 yrs x 365 day/yr)]. Parkins et al. (1974) reported bone fluoride values which were twice those reported by Zipkin et al. (1958) and thus, would be consistent with an increase. These findings have potentially serious implications since an increase in intake could lead to dental fluorosis in fluoridated communities. A retention of 2 mg/day would mean that after 40 years the bone of an average individual would have fluoride levels of approximately 10,000 ppm, i.e., similar to those associated with skeletal fluorosis (Hodge and Smith, 1965). Evaluation of the possibility that an increase has occurred in the intake of fluoride will be considered under three headings: Food, Bone and Urine. Bone should reflect a long-term increase and urine both short-term and long-term increases (Taves and Guy, this monograph).

Food

Kintner (1971) concluded that there is a progressive 4- to 5-fold increase in the amount of fluoride entering the food chain where there is little or no fluoride in the water supply. He quotes Marier and Rose (1966) who analyzed the fluoride content of canned food obtained from canneries using fluoridated or non-fluoridated water. They showed convincingly that there was about 0.8 ppm (42 μM) total

149

fluoride (i.e., after ashing) in products from a cannery
using fluoridated water, and 0.2 to 0.3 ppm in products
from a cannery using non-fluoridated water. They converted
the 0.5 ppm increase in canned samples to a 0.5 mg increase
in daily intake, and then added this to the 0.5 to 1.5 mg
estimated by Hodge and Smith as the daily fluoride intake
from food in a non-fluoridated area (Hodge and Smith, 1965).
This resulted in an estimated range of 1 to 2 mg/day
fluoride intake, from food alone. The Hodge and Smith
estimate was derived from values reported by Machle and
Largent (1943). These values were based, erroneously, on
earlier work by Machle, Scott and Largent (1942) in which
data were given for fluoride intake and on a survey of
urinary values. The data show the average daily fluoride
intake for 20 weeks to be just under 0.5 mg with only
0.155 mg of that being from food, per se. This hardly
supports the 0.5 to 1.5 mg range. In addition, the conversion
from ppm to mg in the Marier and Rose study requires the
assumption of an intake of one kilogram of canned food per
day! Thus, Marier and Rose certainly did not provide
convincing evidence for their estimate of daily fluoride
intake. Hodge and Smith (1970), in a more recent review of
the literature, lowered their estimate to 0.3 to 0.8 mg
F/day from food.

   Two recent articles from Spencer's group appear to
support a higher estimate for dietary fluoride (Kramer et
al., 1974; Osis et al., 1974) than the original estimate of
0.3-0.5 mg/day by McClure (1944). The first of these is
based on hospital-prepared food from 16 U.S. cities. The
fluoride intake from food alone in fluoridated communities
ranged from 1.7 to 3.4 mg per day (average, 2.62 mg),
while the daily intake from non-fluoridated communities was
0.78 to 1.03 mg (average 0.9 mg). The very high values
and the marked difference between F and non-F areas can be
explained in part by inclusion of coffee and other water-
based beverages as part of the diet, which other investigators usually have not done. The second paper reported the
average fluoride intake from diets used in balance studies
in a fluoridated city for a six-year period to be 1.96 mg/
day. No increase in fluoride intake was seen during that
period of time.

   The values found by Spencer's group are suspect
because of the use of Singer and Armstrong's method of
diffusing fluoride at 60°C into a trapping solution and
then measuring it with a colorimetric reagent (Guy, this
monograph). It is clear that something other than fluoride
diffuses and is trapped with this technique since Singer
and Armstrong (1969) reported that fluoride values for
serum are several-fold lower when determined by the fluoride

electrode than those determined with the colorimetric reagent. They also noted that the trapping solutions can be
turbid when the samples are undried food (Singer and Armstrong, 1965). Spencer's group used the colorimetric
reagent and appear to have omitted the recommended drying
step.

A study by San Filippo and Battistone (1971) with a
better method but more limited scope resulted in values of
2.09 to 2.34 mg/day for total fluoride intake. They used
Singer and Armstrong's earlier method in which the food was
ashed before diffusion and measured with the colorimetric
determination of fluoride content. The slight increase in
these values over earlier estimates may be a reflection of
the fact that the food portions were for 16- to 19-year-old
males, i.e., heavy eaters. Data from balance studies in
children support the lower values. The fluoride intake from
the diet of nine children aged 4 to 18 years averaged 0.3 mg/
day (Forbes et al., 1973).

Thus, questions about the average fluoride intake and
balance certainly need further study. Taves and Ning (in
preparation) selected foods which Spencer's group had
reported to be particularly high in fluoride and subjected
them to the same method of separation of fluoride (diffusion
at 60°C). They measured the isolated fluoride by the colorimetric reagent used by Spencer's group and by the fluoride
electrode. The data are shown in Table I.

Note that the discrepancies are as high as 200 fold.
There was good agreement only for the high values in canned
chicken. Taves and Ning show that the discrepancy is not
due to the presence of organic fluoride since no additional
fluoride appeared with combustion of the trapping solution
from diffusion analysis of apple (25-fold discrepancy before
ashing). The conclusion from these experiments is that the
discrepancy is not due to fluoride in any form. (L. Singer,
personal communication, 1977)has found similar discrepancies.

Bone

Charen et al. (1978) have re-analyzed the bone used
by Parkins et al. (1974) and found values which were only
1/4th as high. The repeat analyses were done at two
different laboratories with good agreement. No specific
error could be found in the methodology used by Parkins
et al. so presumably there was some inadvertent technical
error. The fluoride levels of the Iowa bone were not really
suitable for comparison with earlier bone levels since
the hospital records did not give information on how long
the patients had lived in fluoridated communities. A group

TABLE I

Comparison of Measurement Methods
After Diffusion at 60°C

Food	Method	
	Colorimetric μg/gm	Electrode μg/gm
Applesauce and Cherries*	3.25 ±[1] 0.17	0.028 ± 0.002
Creamed Spinach*	2.0 ± 0.08	0.70 ± 0.03
Chicken & Chicken Broth*	5.65 ± 0.17	5.1 ± 0.32
Fresh Apple	4.12 ± 0.45	0.023 ± 0.01
Fresh Cherries	2.60 ± 0.60	0.011 ± 0.01

*Baby Foods

[1] ± S.D., 4 samples

of 17 patients who had lived in a fluoridated community
(Rochester, N.Y.) for at least 7 years was found (Charen
et al., 1978) to have the average bone fluoride concentration
expected on the basis of older data of Zipkin et al. (1958).
These data (Figure 1) suggest that there has been no increase
in fluoride intake in Rochester, N.Y. An increase of
25% can be ruled out on the basis of the standard error, but
both positive and negative potential sources of error make
this estimate uncertain. (A greater proportion of the cases
had renal disease tending to artificially raise the mean,
however, the length of time the subjects might have drunk
fluoridated water was lower in the study of Charen et al.)

Urine

Samuelson et al. (1976) measured the 24-hour urinary
fluoride excretions pre-operatively in 17 patients. They
found an average 1.9 mg fluoride/day. Taves (unpublished
data) collected 24-hour urine specimens for several months
while drinking fluoridated water and eating food prepared
with fluoridated water. These collections were repeated
during a comparable period of drinking distilled water while
using either distilled or tap (fluoridated) water in preparing the food. The results are shown in Fig. 2 as a function
of the total urine volume. The 24-hour urine fluoride
values were between 1 and 2 mg/day with an average of 1.2.
The fluoride excreted when not drinking fluoridated water
was 0.3 to 0.6 mg/day. There was no apparent relationship
between the urine volume and the amount of fluoride excreted
per day.

Leaf and Taves (unpublished data, 1976) measured the
fluoride concentration in spot samples of urine from 43
young (average age, 23) males drinking fluoridated (1 ppm)
water. The results were compared (Figure 3) to values
obtained by McClure and Kinser (1944). Again, the average
fluoride concentration falls close to the expected value.
The standard error of the difference is only 0.038 ppm, so
the probability of the sample being drawn from a population
whose real mean urine fluoride level was 0.81 ppm, or ·
8% above the line drawn through the earlier data, is 1/20,
i.e., a real increase of 10% can be ruled out.

Singer et al. (1969) measured fluoride levels in
urine from six normal humans who lived in a fluoridated city
and presumably drank water with 1 ppm fluoride.
The average urine fluoride was 0.95 ppm $\pm$ 0.17 (S.E.)
which is consistent with the findings of Leaf and Taves.
Again, these data indicate that there has been no change in
fluoride intake when comparing current data to those from two

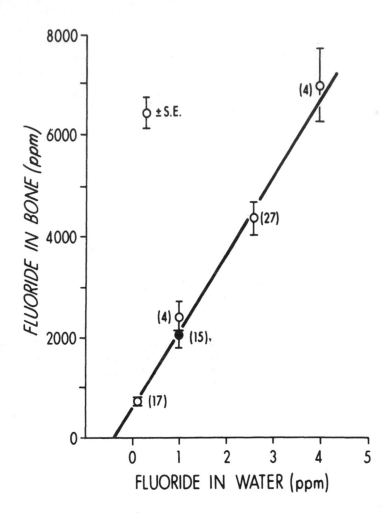

Figure 1

Fluoride concentration in human bone ash versus that in the drinking water, Zipkin et al. (1958).

    0 - Zipkin's subjects

       average age at 0.1 ppm = 57 ± 3.8;
                 1   ppm = 76 ± 1.5;
                2.6 ppm = 66 ± 3.0;
                4   ppm = 56 ±11.

    ● - Rochester subjects, average age 70 ± 3.1

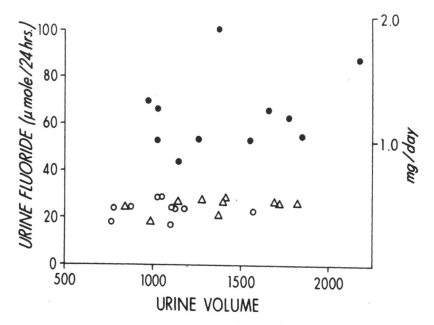

Figure 2. Fluoride excretion per 24 hours as a function of
urine volume in the same subject with different dietary intake,
Taves (unpublished). ● - Fluoridated water for drinking and
cooking; Δ - Fluoridated water for cooking, distilled for
drinking; 0 - Distilled water for drinking and cooking

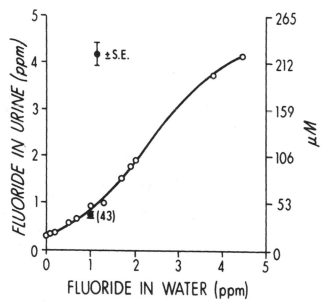

Figure 3. Fluoride concentration in human urine versus that in
the drinking water, McClure et al. (1944). 0 - McClure's subjects; ● - Rochester subjects

decades ago with the subjects' drinking water at the same
fluoride concentration.

Concluding Remarks

It has been shown that evidence of an increase in
fluoride intake is based on faulty analyses. Data obtained
with reliable methods show no indication of an increase in
food, bone or urine concentrations as compared to 20 or more
years ago. The amount of increase in total intake (based
on urine values) that might have occurred and yet have
escaped detection with the available data is less than 10%.

References

Charen, J., Taves, D.R., Stamm, J.W., et al. (1978) Bone
    fluoride concentrations associated with fluoridated
    drinking water. Submitted for publication Calc. Tiss.
    Res.

Forbes, D.B., Smith, F.A., Bryson, M.F. (1973) Effect of
    growth hormone on fluoride balance. Calc. Tiss. Res.
    11: 301-310.

Hodge, H.C., Smith, F.A. (1965) In: Fluorine Chemistry, IV
    (J.H. Simons, ed.). Academic Press, New York.

Hodge, H.C., Smith, F.A. (1970) Minerals: fluorine and
    dental caries. In: Advances in Chemistry Series, No.
    94, pp. 93-115.

Kinter, R.R. (1971) Dietary fluoride in the U.S.A.
    Fluoride 4:2, 44-48.

Kramer, L., Osis, D., Wiatrowski, E., et al. (1974) Dietary
    fluoride in different areas of the United States. Amer.
    J. Clin. Nutr. 24: 590-594.

Machle, W., Scott, E.W., Largent, E.J. (1942) The absorption
    and excretion of fluorides: The normal fluoride bal-
    ance. J. of Indust. Hyg. 24: 190-204.

Machle, W., Largent, E.J. (1943) The absorption and excre-
    tion of fluoride: The metabolism at high levels of
    intake. J. Indust. Hyg. 25: 112-123.

Marier, J.R., Rose, D. (1966) The fluoride content of some
    foods and beverages. J. Food Science 31: 941-946.

McClure, F.J., Kinser, C.A. (1944) Fluoride domestic waters
and systemic effects. II Fluorine content of urine in
relation to fluorine in drinking water. Pub. Health
Rep. 59: 1575-1591.

Osis, D., Kramer, L., Wiatrowski, E., Spencer, H. (1974)
Dietary fluoride intake in man. J. Nutr. 104: 1313-
1318.

Parkins, F.M., Tinanoff, N., Moutinno, M., et al. (1974)
Relationship of human plasma fluoride and bone fluoride
to age. Calc. Tiss. Res. 16: 335-338.

Samuelson, P.N., Merin, R., Taves, D.R., et al. (1976)
Toxicity following methoxyflurane anaesthesia. IV The
role of obesity. The effect of low dose anaesthesia
on fluoride metabolism and renal function. Canad.
Anaesth. Soc. J. 23: 465-479.

San Filippo, F.A., Battistone, G.C. (1971) The fluoride
content of a representative diet of the young adult
male. Clin. Chim. Acta 31: 453-457.

Singer, L., Armstrong, W.D. (1965) Determination of fluoride. Procedure based upon diffusion of hydrogen
fluoride. Anal. Biochem. 10: 495-500.

Singer, L., Armstrong, W.D. (1969) Total fluoride content
of human serum. Arch. Oral Biol. 14: 1343-1347.

Spencer, H., Lewin, I., Wiatrowski, et al. (1970) Fluoride
metabolism in man. Am. J. Med. 49: 807-813.

Taves, D.R., Ning, C.F. Fluoride in food. In preparation.

Zipkin, I., McClure, F.J., Leone, N.C., et al. (1958)
Fluoride deposition in human bones after prolonged
ingestion of fluoride in drinking water. U.S. Pub.
Health Rep. 73: 732-740.

7

Distribution of Fluoride Among Body Compartments

Donald R. Taves and Warren S. Guy

The relationships between the concentration of fluoride in serum and that in other compartments of the body is important in evaluating toxicity and the likelihood of idiosyncratic responses. If there is a simple direct relationship, the margin of safety may be established more precisely from these relationships than can be done from information on fluoride dose alone. For instance, elimination of fluoride decreases in certain renal diseases (Johnson et al., this monograph). Therefore, the toxic dose will be less than that required to cause toxicity in the normal individual. It is difficult to determine whether symptoms are caused by the renal disease itself or by excessive fluoride intake because the signs and symptoms are not specific. If serum fluoride levels were shown to correlate with toxicity, however, then they would provide a means of determining the cause of a patient's symptoms. If, instead of a simple relationship there is a complex regulating system controlling the serum fluoride, the possibility of idiosyncratic response is greater since a complex system has more steps which can fail.

Whether the serum fluoride level does, in fact, have predictive value is still being debated. On the one hand, Singer and Armstrong (1960, 1970) proposed that the serum fluoride concentration is under some sort of homeostatic control, i.e., does not change with conditions as expected, and is therefore not useful in a diagnostic sense. On the other hand, Taves (1970) suggested that the serum fluoride concentration would be useful in monitoring the doses of fluoride used in treating patients with osteoporosis. Ericsson (1973) contended that the serum fluoride concentration is related to the bone fluoride concentration. The compartmental analysis of Hall et al. (1977) also supports

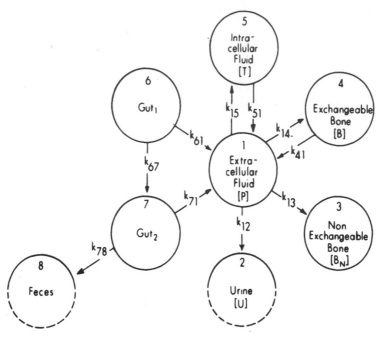

Figure 1. Compartmental model of fluoride metabolism. k_{ij}= fractional rate of transfer into compartment j from compartment i in units of minutes^{-1}. Compartments 6, 7 and 8 represent the gastrointestinal tract; compartments 2 and 3 are irreversible compartments and compartments 1, 4 and 5 are exchanging compartments. From Hall et al. (1977).

TABLE 1

Fractional Transfer Rates[a]

	Per minute	
	This study	Data of Costeas (1968)
k_{12}	0.00621	0.00822
k_{13}	0.0210	0.0180
k_{31}	0.0000154	--
k_{14}	0.00596	0.0120
k_{41}	0.00274	0.00915
k_{15}	0.0422	0.0184
k_{51}	0.0911	0.0773
k_{61}	0.0285	na
k_{67}	0.0855	na
k_{71}	0.0151	na
k_{78}	0.0000	na

[a] Note: na = nonapplicable. From Hall et al. (1977)

this view. Since homeostasis of serum fluoride has been the subject of debate without appropriate definition of terms a detailed consideration is needed.

This review will first consider compartmental analysis of fluoride, and the concept of homeostasis and then discuss data showing the relationships between serum fluoride and fluoride concentrations in other compartments. Following that, evidence for a relationship of serum fluoride to age and to toxicity will be considered. The terms serum fluoride and plasma fluoride are used here interchangeably since no difference has been noted between the two concentrations and some of the reported work has been done on each.

Compartmental Analysis

Compartmental analysis involves the selection of as simple a mathemathical model of the behavior of a substance as is consistent with its known behavior. As the movement of chemicals is generally passive (down chemical gradients); this is presumed to be the case until proven otherwise. The compartmental model proposed by Hall et al. for the rabbit is shown in Figure 1. This model is based on plasma and urine data obtained after administration of fluoride to rabbits. The concentration of plasma fluoride ranged from 100 to 1 μM. The model shows fluoride entering from two distinct gut compartments. Since pH has a bearing on the absorption of weak acids like HF (Whitford et al., 1976), the two gut compartments may represent stomach and intestine where the pH is approximately 1 and 7.5, respectively. The model shows fluoride moving from the gut into a central compartment, the extracellular fluid, and from there into two compartments associated with bone, as well as into the intracellular fluid and the urine. Table 1 shows that the relationships between the various compartments can all be adequately described by rate constants, rather than concentration-dependent transport coefficients. This implies that regardless of the concentration of fluoride in any compartment (within the limits studied) the values of the rate coefficients remained constant. For example, if the amount absorbed by the gut per day was doubled, the steady state concentration in every other compartment would double. This fundamental relationship, if generally true, would be of great importance because one would need to measure a concentration in only one compartment in order to know the concentration present in every other compartment. The only condition needing to be met would be that a steady state condition (equilibrium) had been reached. Serum is a central compartment to which all compartments (except fecal) are attached. Thus, the relationship of the serum fluoride concentration to that of any one compartment would be independent and

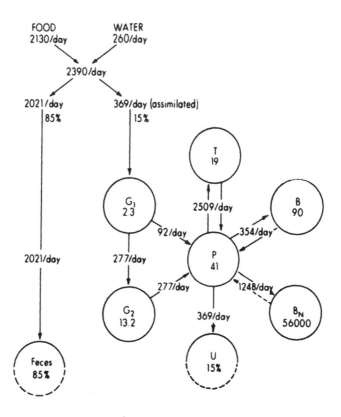

Figure 2. Predicted steady state distribution of dietary
fluoride in the rabbit. The mass in all fluxes and in the
compartmental contents is expressed in micrograms. From
Hall et al. (1977).

undisturbed by diseases which change the rate constant for any other compartment.

The size of the fluoride pool in the compartment plays a major role in determining its influence on the serum fluoride concentration. Figure 2 shows the sizes of the compartments as estimated by Hall et al. It shows, as is well known, that the bone contains most of the fluoride in the body, in this case, of the rabbit. It contains 99.7% of the fluoride in the tissues of the body or 96%, if the nonabsorbed fluoride in the food in the gastrointestinal tract is included. Fluoride enters and leaves the large bone compartments, causing fluctuations of the concentrations in other compartments to be dampened or buffered. The exchangeable bone compartment buffers short term changes (hours) while the long term changes (days to years) would be buffered by the larger "non-exchangeable" compartment as the bone is undergoing remodeling. Hall et al. estimated the half-time for the long term (non-exchangeable) compartment to be 137 days. This means (in the rabbit) that concentration of fluoride in the bone compartment will come to within 99% of equilibrium with the other compartments in 685 days (137 x 5) or two years after a change in the daily intake has occurred. They estimated that the bone compartments determine 80% of the concentrations in the other compartments. This means that where the daily intake has been constant for more than two years, the serum fluoride concentration should change no more than 20% during 24 hours. If the daily intake of fluoride is not absorbed in a short time (minutes) even this 20% figure is reduced. It becomes zero if the absorption is uniform throughout the 24 hours. Thus, the daily fluctuations of the serum fluoride concentrations will be small because of the buffering action of the bone compartment. When equilibrium throughout the whole system is achieved, the serum fluoride concentration will be directly proportional to the average daily intake for this model.

Evidence for homeostasis of serum fluoride

The term "homeostasis" in respect to fluoride has been used without sharp definition and without regard to the concepts derived from the above model. Dorland's Medical Dictionary up through 1965 (24th ed.) defines homeostasis as "a tendency to uniformity or stability in the normal body states (internal environment or fluid matrix) of the organism (Cannon)". The dampening of changes in serum fluoride concentration predicted by the simple model of Hall et al. is, therefore, homeostasis by that definition. The new definition, Dorland's Medical Dictionary, 25th edition, is significantly different: "a tendency to stability in the normal body states (internal environment) of the organism. It is achieved by a system of control mechanisms activated by negative feed back; e.g., a high level of carbon dioxide in extracellular fluid triggers increased pulmonary ventilation, which in turn causes a decrease in carbon dioxide concentration." Singer and Armstrong (1960, 1970) use the term homeostasis interchangeably with regulation and suggest the serum fluoride is maintained independently of the intake and bone fluoride levels. This would be consistent with the modern definition. However, they have not shown or postulated the presence of the negative feedback systems that would be necessary to achieve homeostasis in the modern sense. When used in the old or vague sense it will be used here with quotation marks. Homeostasis (regulation) would require a complex model, one in which the transport rate coefficient varied as a function of the concentration of fluoride.

The term "homeostasis" was first used by Singer and Armstrong in reporting a study in which they measured plasma fluoride concentrations for five groups of humans who had been drinking water with fluoride content ranging from 0.15 to 5.4 ppm F for at least three years. They found no difference, using an ashing-distillation-colorimetry method, between the average serum fluoride values among four groups drinking water containing from 0.15 to 2.5 ppm F (Guy, this monograph, Table 1, Section E2). They reported similar findings in rats receiving 80 ppm F in food and 0 to 50 ppm F in water (Singer and Armstrong, 1964). From these observations they concluded that regulatory mechanisms operate within the body to maintain plasma fluoride within narrow limit with substantially varied intake. In their 1970 review, Armstrong and Singer considered the following additional points in support of their hypothesis: (1) Carlson et al. (1960a) found only transient (hours) and small (equivalent to 8-10% of dose) rises in plasma fluoride levels following ingestion of a one-milligram dose labeled with ^{18}F; (2) Rich et al. (1964) found small relative increases of plasma fluoride in patients with metabolic bone disease who received 50-100 mg F per day and reported that the increase diminished with continuation of the same dose; (3) the average blood fluoride reported by Smith et al. (1950) showed only a three-fold increase in two human populations drinking water with a twenty-three-fold difference in fluoride content; (4) Smith et al. (1960) found little difference in tissue fluoride levels in corpses from communities where the fluoride content of the drinking water varied from 0 to 4 ppm; (5) Call et al. (1965) found no increase in fluoride levels in tissues of corpses exposed to high atmospheric fluoride pollution; (6) Singer et al. found no alteration of plasma fluoride levels in rats raised on high fluoride intake when treated with parathyroid hormone or when starved (1964) or nephrectomized (1965), and

(7) Shupe et al. (1963) reported the same range of blood fluoride in cattle with large variations in intake and urine fluoride.

Singer and Armstrong's original data on humans and rats were obtained with ashed samples and are therefore unreliable because they represent total fluorine rather than inorganic fluoride (Guy, this monograph). Taking the balance of the supporting evidence point by point; first, the rapid disposal of plasma fluoride following ingestion of a single small dose is not questioned here. With soft-tissue and extra-vascular compartments roughly 7 times the size of the intravascular space (Figure 2), no more than 15% of the dose would be expected to appear in the plasma, assuming more rapid transfer from serum to those compartments than from gut to serum. Thus, their observations are expected from the simple model.

The second report given to support the homeostasis theory was described as follows, "The plasma fluoride levels of patients with metabolic bone diseases treated by Rich, Ensinck and Ivanovich (1964) with large doses of sodium fluoride (50-100 mg fluoride per day) for 10-34 weeks did not rise above 1.8 ppm. The one patient whose plasma fluoride level reached 1.8 ppm on the 14th day of treatment, from the pretreatment level of 0.32 ppm, continued to receive 50 mg fluoride daily and, after 33 and 34 weeks of treatment, the plasma fluoride content had declined to 0.45 and 0.48 ppm" (Armstrong et al., 1964). Apparently due to an oversight, it was not mentioned that on the 14th day of treatment this patient began taking aluminum hydroxide gel (twice daily) which, according to Rich et al. led to very poor fluoride absorption. The decrease may also have been due to increased bone formation (the purpose of the treatment) and clearance of fluoride by bone. The magnitude of the serum fluoride elevation relative to the expected pretreatment inorganic fluoride concentration (they give a total value) represents a 75-fold increase (0.02 to 1.5 ppm). The expected increase in this case where the dose was 100 mg F rather than the 2 mg normally ingested (Taves, this monograph) would be 50-fold assuming either that equilibrium had been reached with the bone or that the peak serum fluoride concentrations were measured both before and during treatments. The authors indicate that the blood samples were drawn at fasting and that treatment with fluoride had been carried out for less than one year, therefore, neither assumption is met and the level of 1.8 ppm (90 μM) fluoride is surprisingly high. The data thus can be explained by a simple model, and the concept of homeostasis is not needed.

With respect to the third point, reference is made to the values of Smith et al. in Section D4 in Table 1 (Guy, this monograph). The argument is that the average values (including zero values) showed only a three-fold variation in blood fluoride concentrations for a twenty-three-fold variation in fluoride intake from water. This comparison is misleading because the total fluoride intake is not considered. Assuming an equal intake of 0.5 mg per day from other sources for each group and an average water intake of one liter per day, the total intakes for the two groups would be 0.6 and 1.9 mg per day, respectively. This amounts to a three-fold difference in intake for the observed three-fold difference in blood fluoride concentrations. Thus, instead of being evidence for homeostatic control, the data are evidence for the simple model.

The fourth and fifth points are based on tissue rather than serum concentrations and thus are not compelling arguments for the hypothesis of serum fluoride homeostasis. These data will be considered separately under the heading of tissue levels.

The sixth point, lack of response of the serum fluoride to parathyroid hormone or starvation and little change with dose, was based on a study on rats with fluoride determined after ashing i.e., total fluorine rather than inorganic fluoride (Singer and Armstrong, 1964). The analytical method used for the study of fluoride in the nephrectomized rats is not given (Singer et al., 1965). The results are questionable because the nephrectomized animals ate and drank very little compared to the sham-operated controls. Also, the experiment lasted only 48 hours, a time period for which the simple model would predict little change in serum level since the bone compartments would not have changed much. Experiments using 5/6 nephrectomized rats which survived several months showed a 2 to 3-fold increase in bone fluoride. They showed a several-fold increase in serum fluoride analyzed after low temperature diffusion with the morin-thorium reagent (see discussion of that method in Guy, this monograph). A 3-fold increase was observed a few weeks prior to death, and there was a 7-fold increase over the controls just prior to death (Taves and Morrison, 1970). These results are consistent with the simple model.

Taking the last point, Armstrong and Singer (1970) state that "the detailed studies of Shupe et al. (1963) demonstrated the ability of cattle to regulate the fluoride level of the blood even when the dietary intakes of fluoride were such that various degrees of fluorosis were present." On the contrary, Shupe et al. reported average values for fluoride in cow blood nearly proportional to the concentrations in the feed. For animals on diets containing 12, 27, 49 and 93 ppm F for 7.3 years, they reported average blood levels of 0.07, 0.11, 0.21 and 0.45 ppm F, respectively.

In summary, none of the points raised in support of the concept of "homeostasis" are compelling. Either the methods were unreliable, the evidence not germane or, in fact, the data actually supported the opposite conclusion. The simple model of Hall et al. adequately explains the behavior of fluoride in serum.

Relationship of the Fluoride Concentration in Serum to those in Other Compartments

As discussed above, the evidence for homeostatic control of serum fluoride levels is faulty. Below is a review of additional evidence which supports the simple model, i.e., a direct relationship of serum fluoride to intake, bone, urine, milk, salivary, placental and cord serum levels. Tissue fluoride concentrations which do not always fit the simple model are also reviewed and discussed.

Serum vs. Intake Levels

There is some diurnal variation in the serum fluoride which should be considered first when relating the serum fluoride concentration to long-term intake. The increase over a fasting baseline can be appreciable but is probably not more than 30% as a general rule. Table 2 shows fasting values for serum fluoride and the peak values reached about 50 minutes after fluoride ingestion for two subjects who drank 0.5 mg fluoride in 500 ml water within a minute (Post and Taves, unpublished). Neither subject showed more than a doubling in serum fluoride for this unusually rapid and large intake. Ekstrand (1977a) took blood samples every 4 hours for 36 hours from five subjects whose drinking water contained 1.2 ppm. Three of the 5 increased 10% to 20% and the maximum was a doubling.

However, large differences in plasma fluoride concentration occur with different long-term intakes (Guy et al., 1976). Average plasma fluoride concentrations (blood-bank specimens) from three Texas and two New York cities are shown in Figure 3 as a function of the fluoride concentration of the water supply. A linear relationship between serum and water concentrations is seen with a slope of 0.015. Extrapolation to zero plasma fluoride would give a drinking water value of minus 0.3 ppm, indicating that approximately 0.3-0.5 mg fluoride a day comes from food (assuming 1 to 1 1/2 liter of water consumption per day). This is in good agreement with the values reported in the chapter on fluoride intake from food (Taves, this monograph). There are seven other studies in which reliable methods were used and similar increases in serum F were found (Guy, this monograph, Table 1; Smith et al., 1950; Cox and Backer-Dirks, 1968;

TABLE 2

Observed and Expected Serum Fluoride Concentrations (μM)
After Ingestion of 500 ml fluoridated water.

| | Subject | |
	A.P.	D.R.T.
Serum fluoride after 10–12 hour fast	0.59*	0.74
Serum fluoride after ingestion of 25 μmoles F	1.28	1.36
Expected* increase with no clearance	0.78	0.65
Minus 50 ml/min kidney clearance	0.72	0.60
Minus 150 ml/min bone clearance of excess	0.65	0.57
Actual increase	0.69	0.62

*The 26 μmoles diluted into extracellular water and 60% of
intracellular water.

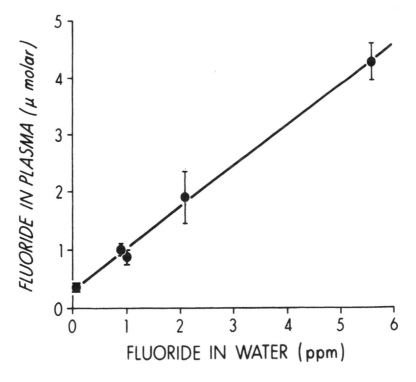

Figure 3. Fluoride concentration in human plasma vs that in
the community water supply. Note the close correspondence
of the numbers when the values are expressed as μM for serum
and as ppm for water.

Jardillier and Desmet, 1973; Inkovaara et al., 1973;
Hanhijärvi, 1975; Ekstrand, 1977a; Hellstrom, 1976).

The same relationship holds true for animals other than
man. In rats the serum fluoride concentration has been related to that of the water supply (100 days) over a much
wider range as shown in Figure 4 (Taves and Raisz, unpublished data). This experiment showed a considerable break in the
curve at 7.5 mM F (150 ppm). Although the reason for this
break is still not clear, the possibility that toxicity had
altered the pattern of intake was checked. Suttie (1968)
reported that rats given toxic concentrations of fluoride in
food ate about four times more food during the daytime when
rats normally fast. We were, therefore, concerned that our
toxic animals may have consumed more water and thus fluoride
just prior to sacrifice. Consequently, another group of rats
was given drinking water with similar fluoride concentration
for 6 and 12 weeks, but the fluoridated water was replaced by
distilled water for 16 hours prior to sacrifice (Au and Taves,
unpublished data). The data from 6 and 12 weeks were not dis-
tinguishable and are plotted together (Figure 5). The animals in this second experiment were fed a special low-fluoride diet (Taylor et al., 1961) which contained less than 1
ppm fluoride while the first experiment used commercial rat
chow which contained 30 ppm fluoride. The relationship between serum fluoride and water fluoride shown in Figure 5 is
more nearly linear for the second experiment. This, however,
is due to a higher value at 7.5 mM rather than to a lower
value at 10 mM. Thus, a change in the pattern of water intake cannot be the full explanation for the break in Figure 4.
The slope of the serum vs water fluoride concentration is
0.003 for the initial portion of the curve for both experiments.

Serum vs. Bone Levels

Bone vs drinking water fluoride concentrations are
shown in Figure 1 of the preceeding chapter of this monograph. Since the fluoride concentrations of serum and bone
are each linearly related to the concentration in the water
supply each must also be linearly related to the other. This
relationship can be described as a ratio which holds true as
a first approximation for any level of steady daily intake.
Combining the data for humans from the two figures gives a
ratio between bone and serum of 100,000 to 1 with fluoride
content expressed per gram of bone ash and per ml of serum.
Bone fluoride concentrations for rats corresponding to the
serum fluoride concentrations in Figure 4 are shown in Figure
6. The two experiments resulted in somewhat different ratios
(see legend). Using the data from the second experiment
(the serum fluoride concentrations were more nearly linear
over the whole range of doses) gives a ratio of 26 micro-
moles per gram bone ash to 1 nmole per milliliter serum or

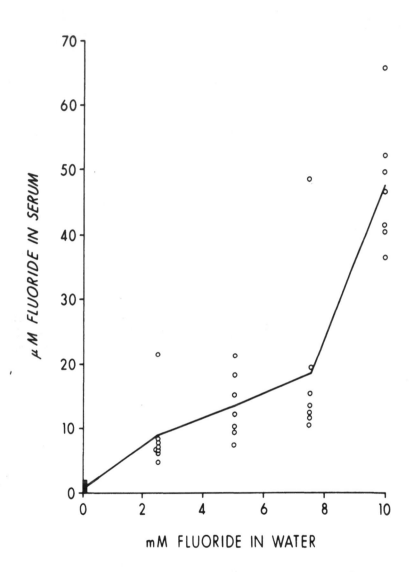

Figure 4. Fluoride concentration in rat serum vs that in the drinking water. The duration of the exposure was 100 days and the diet was a commercial rat chow with 30 ppm. The average serum fluoride concentration with distilled water was 1.2±0.2 μM (±S.E.M.)

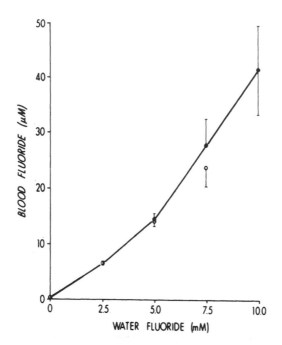

Figure 5. Fluoride concentration in rat serum vs that in drinking water. The duration was 6 and 12 weeks and the diet was a low-fluoride rat chow. The serum fluoride concentrations with distilled water were all less than 0.3 µM at 6 weeks and 0.32±0.03 µM (±S.E.M.) at 12 weeks.

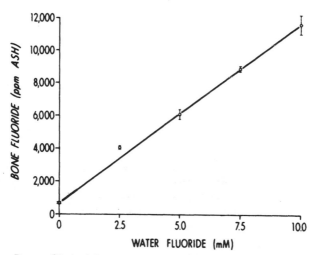

Figure 6. Bone fluoride concentration in rats corresponding to Fig. 4. These values for Fig. 5 describe a line with a slope of 1,750 ppm per mM in the water instead of 1000 shown here.

a ratio of about 25,000 to 1. It is not clear at this time whether the difference between the rat and the human is due to differences in exposure time or to the difference in bone structure (Haversian vs non-Haversian).

Serum vs. Urine Levels

The relationship between fluoride in urine and that in the water supply as found by Zipkin et al. (1956) is shown in Figure 3 of the preceeding chapter of this monograph. It is not a very straight line for reasons which are not clear so the study needs to be repeated. However, the relationship between urinary and drinking water fluoride concentrations was considered to be linear by Hodge et al. (1970) with a ratio of roughly 1. As with bone, since urine and serum fluoride concentrations are both related to that in the drinking water, they have to be related to each other. The ratio of serum to drinking water or to urine is nearly unity when the serum concentration is expressed as μmolar (μM) and the urine or water supply concentration is expressed as ppm, or roughly 0.02 when all are expressed in the same units. Since the average urine flow rate is approximately 1 ml per minute, the renal clearance rate (ml of blood cleared of F per minute) expresses the inverse of this ratio and is more precise than the urine concentration. Unfortunately, few clearance data are available and they are not in good agreement at low fluoride concentrations. Hanhijärvi (1975) found an average clearance rate of 40 ml/min in subjects drinking 1 ppm water and a rate of only half that with 0.2 ppm F drinking water. Ekstrand (1977a) found renal clearance rates of 50.4, 53.5 and 35.9 for individuals drinking water at 9.1, 1.2 and 0.25 ppm F which clearly do not agree with Hanhijärvi's data at the lower intakes. The low clearances probably arise from a systematic over-estimation of the serum fluoride concentrations for those individuals drinking water at the lowest fluoride concentration. The fluoride electrode was used directly on serum in a concentration range which others have found unreliable (Singer and Armstrong, 1973; Fry and Taves, 1970; Guy, this monograph).

The renal clearance may not be a constant and might provide a degree of long-term homeostasis of a type different from that suggested by Singer and Armstrong. If fluoride clearance increases with increasing urine flow rates (i.e., increasing consumption of water), the increase in serum fluoride would not correspond to the increase in the amount of fluoride ingested. Thus, the serum fluoride concentration would be related more closely to the concentration of fluoride in the drinking water than to the amount of fluoride ingested with the water.

Unfortunately, the data on the effect of urine flow rate on renal clearance are not clear as yet. Walser et al. (1966) and Carlson et al. (1960a) found a small increase in fluoride clearance with increasing urine flow rates. Whitford et al. (this monograph) found the same relationship but do not agree that urine flow rate, per se, is the causative variable. They think that the observed change in fluoride clearance is due to an increase in pH resulting from dilution of the filtrate. The effect noted by these three groups of investigators, however, was with high urine flow rates. Renal clearance as a function of urine volume in the range which is encountered normally was reported by Ekstrand and is shown in Figure 7. On the other hand, the urine data shown in Figure 2 of the preceeding chapter of this monograph indicate no increase in total fluoride output per day with increasing volumes of urine. The later data, however, are not solid evidence without serum fluoride concentrations to show that the clearances were also constant.

Serum vs. Milk Levels

The fluoride concentration in human milk is important in that it represents a natural standard or reference point for evaluating fluoride intake and fluoride supplement levels for infants who drink bottled milk. Four recent attempts have been made to analyze breast milk for fluoride. Simpson and Tuba (1968), used the fluoride electrode directly after buffering, and deproteinating the milk and found an average concentration of 25 μM F in the milk of 82 women drinking fluoridated water. This value is hard to believe as it is about 25 times the normal plasma level and is much higher than the milk levels reported by the other three investigators. Scrutiny of the method reveals that acetic acid, NaOH and water were added to the samples which were also manipulated in several ways including boiling and centrifugation in glassware. No mention of blank determinations was made. Ericsson (1969) analyzed human milk with an ashing-diffusion-colorimetric method. He found an average of 1.2 μM F in 11 assays or roughly what one would expect the plasma to be. Konikoff (1974) found the fluoride activity to be below the level of linear response of the electrode. Backer-Dirks et al. (1974) found averages of 0.2 and 0.4 μM F in milk from 10 and 9 mothers drinking water containing 0.1 and 1.0 ppm F, respectively. These values were obtained by direct measurement with the fluoride electrode even though they were below the level of linear response. Thus, even though the analyses of human milk are not ideal, it would appear that the fluoride concentration is quite similar to that in plasma. The corresponding daily fluoride intake for an infant on breast milk would be less than 0.02 mg, which

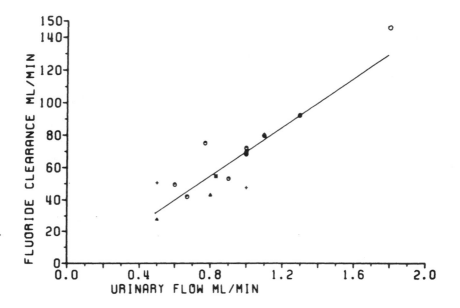

Figure 7. The dependence of renal fluoride clearance on urinary flow in 16 per os and intravenous experiments. 2.82 mg per os and 3 mg intravenously (•), 4.5 mg per os (+), 5.64 mg per os (✗), 9.40 mg per os (▲). The correlation is significant (r=0.90; p < 0.001, n=16). From Ekstrand, Jan, Mats Ehrnebo and Lars O. Boreus: Fluoride bioavailibility after intravenous and oral administration: Importance of renal clearance and urine flow. Clin Pharmacol. Ther. 23: 329–337, 1978.

is only 4% of the presently recommended supplement for infants.

Ashing of milk prior to analysis (Backer-Dirks et al., 1974) gave values 6-12 times higher than those obtained with unashed samples, but without information as to whether blanks with comparable amounts of Ca and PO_4 were used, these data may not be reliable. Milk contains Ca salts which would be capable of fixing any fluoride contamination present in furnace air during the ashing process, while blanks without a fixative would not. Therefore, the high ashed values may be an artifact.

Serum vs. Saliva Levels

Saliva (duct) has been analyzed for fluoride directly and by known-addition with the fluoride electrode has been found to contain a concentration comparable to that in plasma (Grøn et al., 1968). Others using the same method report similar results, namely, fluoride concentration in saliva between 0.4 and 1.6 µM (Ericsson, 1969; Yao and Grøn, 1970; Shannon et al., 1973). Data from Ekstrand (1977b) show a high degree of correlation of serum with salivary fluoride over a range of 1.0 to 7.3 µM, Figure 8.

Serum vs. Placental and Cord Serum Levels

There is only one study on the relationship of maternal to cord blood in which the analytical method would have been specific for inorganic fluoride (Shen and Taves, 1974). Figure 9 shows the average values obtained for maternal and cord blood to be, 0.88 and 0.68 µM F, respectively. A strong positive correlation between maternal and fetal blood levels was apparent in the 16 sets reported. These data, do not disprove the barrier hypothesis, however, because the slowing of cord-blood flow after delivery may have allowed the fluoride to approach equilibrium across the placenta.

A placental barrier to fluoride has been postulated which might protect the infant (Gedalia, 1970). However, finding a fetal blood level lower than the maternal blood level would not prove that there was more than a passive diffusion barrier.

Reasoning from ^{18}F data is particularly tricky for determining the relationship of maternal blood to cord blood. The amount of fluoride found in the placenta at birth is about 1 mg (50 µmoles) based on the concentrations found by Gardner et al. (1952) and Shen and Taves (1974) and could have been supplied by blood with 1 µM fluoride flowing at 0.5 L/min (Williams Obstetrics, 1976) in 100 minutes. This period is long when compared to the time span that radioactive fluoride would be presented to the placenta after an

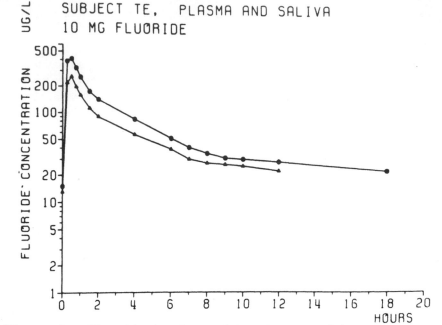

Figure 8. Fluoride in plasma (●) and saliva (▲) versus time after a single oral dose of 10 mg F as NaF tablets. From Ekstrand (1977b).

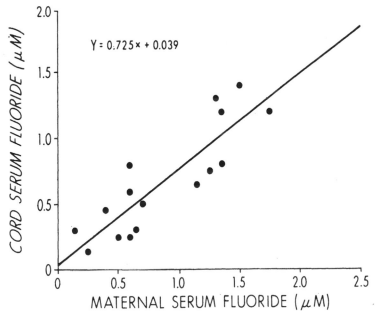

Figure 9. Scattergram of the maternal versus the cord serum fluoride values taken at the time of delivery. From Shen, Y.W. and Donald R. Taves: Fluoride concentrations in the human placenta and maternal and cord blood. Am. J. Obstet. Gynecol. *119:* 205–207, 1974.

intravenous injection. Thus, uptake of radioactive fluoride by placental calcification may be the explanation for the low concentration of radioactive fluoride found in cord blood. However, a 100-minute time period is insignificant from the point of view of stable fluoride.

Serum vs. Tissue Levels

Tissue fluoride concentrations which have been determined with reliable techniques are scarce and the data appear to be in conflict. Studies with ^{18}F generally show lower levels in most soft tissues than in plasma. This is most apparent when data are expressed as the equilibrium ratio (cpm in tissue water:cpm in plasma water) as given by Carlson et al. (1960b) for normal rats. They reported mean ratios as 0.9, 0.8, 0.7, 0.5, 0.5 and 0.2 for skin, testes, liver, muscle, heart and brain, respectively. Similar values are given by Wallace-Durbin (1954) and Hein et al. (1956). The liver vs serum fluoride ratio for the groups of rats in Figure 4 ranged from 0.70 to 0.87 (Taves and Raisz, unpublished). Acceptance of the low-range values for fluoride in plasma (Guy, this monograph) poses a problem with the high values obtained by measurement of fluoride in autopsy tissues. The values obtained by Smith et al. (1960) are by the distillation-ashing-distillation-colorimetry method shown to be relatively specific for fluoride in blood (Guy, this monograph). Figures 10 and 11 summarize the data of Smith et al. (1950) for autopsy tissues (liver and heart) as a function of the concentration of fluoride in water. There is clearly no indication of a correlation between tissue and drinking water concentrations. Furthermore, the tissue concentrations are 25 to 60 times greater than the expected serum level with 1 ppm F drinking water. High autopsy tissue levels have been confirmed with the fluoride electrode after separation by the diffusion method (Taves, unpublished). The best tentative explanation of the high tissue fluoride concentrations in human autopsy specimens is that fluoride is sequestered by calcium phosphate precipitates just prior to death. The lack of increase in tissue fluoride with increasing fluoride content in water, however, needs explanation. The explanation may be that higher serum fluoride concentrations inhibit the precipitation of calcium phosphate (Taves and Neuman, 1964) making the fractional uptake of fluoride less. The concentration of fluoride in tissue relative to that in serum(25- to 60-fold higher) requires that sequestration of fluoride must be occurring for at least an hour prior to death.

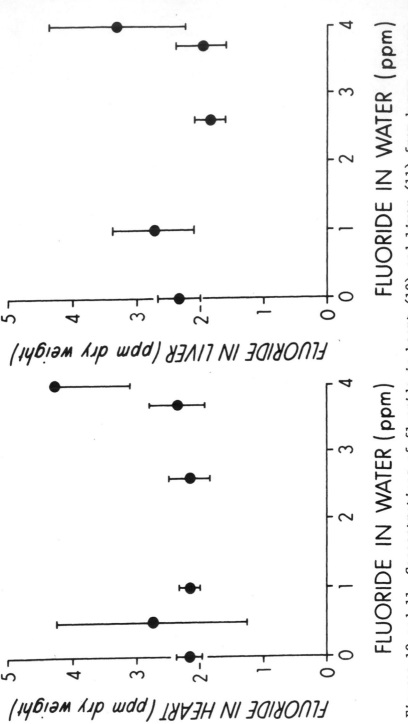

Figures 10 and 11. Concentrations of fluoride in heart (10) and liver (11) from human corpses as a function of the fluoride concentration in the water supply. From Smith et al. (1960).

Relationship of Serum Fluoride to Age

Serum fluoride levels also vary with age. The bone fluoride concentration in humans has been shown to increase until the age of about 50 years (Hodge and Smith, 1965); therefore, a corresponding increase in the serum fluoride concentration would be expected according to the simple model. An increase has been reported by Parkins et al. (1974); Kuo and Stamm (1975); Hanhijärvi (1975) and Husdan et al. (1976). The change in serum fluoride noted with age is not very great and the variation in the reports is considerable, probably due to the unreliability of direct readings with the fluoride electrode below its range of linear response. It is, therefore, likely that the relationship will be made clearer with more accurate analytical methods.

Relationship of Serum Fluoride to Toxicity

Knowledge of plasma fluoride levels associated with various toxic responses to fluoride is important both clinically and in the study of mechanisms of toxicity as discussed in the introduction. Plasma fluoride levels relating to the production of dental fluorosis can only be estimated, at the present time. Average plasma fluoride levels for adults living in areas with 2.1 and 5.6 ppm F in the drinking water were reported to be 1.9 and 4.2 μM, respectively (Figure 3). The higher water fluoride level clearly produces an undesirable degree of dental fluorosis. It is likely that the peak plasma fluoride level following ingestion plays a more significant role in the development of dental fluorosis than does the steady-state level (Angmar-Mansson et al., 1976). Ingestion of 200 ml of 5.6 ppm F water by a 10 kg child would result in a peak plasma fluoride concentration of about 10 μM assuming that 10% of the ingested dose is the maximum found in plasma (Carlson et al., 1960a).

Singla et al. (1976) reported that the concentration of serum fluoride associated with skeletal fluorosis averaged 8.76 μM F in ten patients. They used the fluoride electrode after diffusion with hexamethyldisiloxane. The values may not be representative of the whole year since the intake of water would be expected to be greater in the hot season in India which is not the usual time for research work in the field (Jolly, S.S., personal communication). Serum concentrations in patients receiving single ten-milligram oral doses of fluoride reached a peak of about 20 μM (Ekstrand, 1978) and diminished to less than 10 μM and often less than 5 μM by the next morning even when given daily for long periods of time (Jowsey, this monograph).

Serum fluoride levels associated with renal toxicity have been studied in some detail because of the finding that methoxyflurane is extensively metabolized, releasing inorganic fluoride which causes high-output renal failure. The serum fluoride concentration necessary to produce noticeably increased urine flow rates is about 50 μM in the rat (Whitford and Taves, 1971, 1973; Mazze et al., 1972) and in the human (Mazze, 1971a and 1971b).

The serum fluoride level associated with loss of weight on a chronic basis in the growing rat is about 15 μM (Raisz and Taves, 1967; Figures 4 and 6). The animals with concentrations of 40 to 50 μM F were obviously intoxicated and had increased mortality rates (Taves and Raisz, unpublished data; Au and Taves, unpublished data). Acute lethal concentrations in the rat were 500-1000 μM (de Lopez et al., 1976). A human has survived 500 μM with hemodialysis (Berman et al., 1973).

Concluding Remarks

The data and analysis presented above indicate that the serum fluoride level is a good steady-state predictor of the concentrations in other compartments of the body except for soft tissues of autopsy cases and the placenta at term. The need for such a predictor arises when the long-term intake has changed or is uncertain and when there is a disturbance of the normal relationship between intake and the various compartments. The daily fluctuations of fluoride in blood are usually held to a narrow range of about 30% change by the dampening action of the very large bone compartment. The urinary clearance of fluoride (Johnson et al., this monograph; Taves and Morrison, 1970) can be greatly reduced by renal disease so that the normal relationship between intake and bone is altered, making the serum fluoride the only easily obtained predictor of what is happening in the bone. The ability of the simple compartment model to describe changes occurring in serum fluoride under various condition means that there are probably no complex regulating systems which may fail for some individuals and cause idiosyncratic responses. The effect of renal insufficiency, of course, needs more consideration (Johnson et al., this monograph) and the explanation for the lack of dependence of the tissue fluoride on estimated intake needs verification.

References

Angmar-Mansson, B., Ericsson, Y. and Ekberg, O. (1976) Plasma fluoride and enamel fluorosis. Calc. Tiss. Res., 22:77-84.

Armstrong, W.D. et al. (1964) Plasma fluoride concentrations of patients treated with sodium fluoride. J. Clin. Invest., 43: 556.

Armstrong, W.D. and Singer, L. (1970) Distribution of fluoride in body fluids and soft tissues. In Fluorides and Human Health, WHO Monograph, series 59, Geneva, pp. 94-104.

Backer-Dirks, O. et al. (1974) Total and free ionic fluoride in human and cow's milk as determined by gas-liquid chromatography and the fluoride electrode. Caries Res., 8; 181-186.

Berman, L.B. et al. (1973) Inorganic fluoride poisoning: Treatment by hemodialysis. N.E.J. Med., 289; 922.

Call, R.A. et al. (1965) Histological and chemical studies in man on effects of fluoride. Pub. Health Rep., 80; 529-538.

Carlson, C.H., Armstrong, W.D. and Singer, L. (1960a) Distribution and excretion of radiofluoride in the human. Proc. Soc. Exp. Biol. Med., 104; 235-239.

Carlson, C.H., Singer, L. and Armstrong, W.D. (1960b) Radiofluoride distribution in tissues of normal and nephrectomized rats. Proc. Soc. Exp. Biol. Med., 103; 418-420.

Cox, F.H. and Backer-Dirks, O. (1968) The determination of fluoride in blood serum. Caries Res., 2; 69-78.

deLopez, O.H., Smith, F.A. and Hodge, H.C. (1976) Plasma fluoride concentrations in rats acutely poisoned with sodium fluoride. Tox. Appl. Pharm., 37; 75-83.

Dorland's Medical Dictionary, 24th & 25th ed., W.B. Saunders, Philadelphia, 1965, 1974.

Ekstrand, J. (1977a) Studies on the pharmacokinetics of fluoride in man. Norstedts Tryckevi, Stockholm.

Ekstrand, J. (1977b) A micromethod for the determination of fluoride in blood plasma and saliva. Calcif. Tiss. Res., 23; 225-228.

Ekstrand, J., Ehrnebo, M., Boreus, L.O. (1978) Fluoride bioavailability after intravenous and oral administration: Importance of renal clearance and urine flow. Clin. Pharmacol. Ther., 23; 329-337.

Ericsson, Y. (1969) Fluoride excretion in human saliva and milk. Caries Res., 3; 159-166.

Ericsson, Y., Gydell, K. and Hammarskiöld, T. (1973) Blood plasma fluoride: An indicator of skeletal fluoride content. Internat'l. Res. Comm. Syst., L; 33.

Fry, B.W. and Taves, D.R. (1970) Serum fluoride analysis with the fluoride electrode. J. Lab. Clin. Med., 75; 1020-1025.

Gardner, D.E. et al. (1952) The fluoride content of placental tissue as related to the fluoride content of drinking water. Sci., 115; 208-209.

Gedalia, I. (1970) Distribution in placenta and factors.   In Fluorides and Human Health, WHO Monograph, series 59, Geneva, pp. 128-134.

Grøn, P., McCann, H.G. and Brudevold, F. (1968) The direct determination of fluoride in human saliva by a fluoride electrode. Archs. Oral Biol., 13; 203-213.

Guy, W.S., Taves, D.R. and Brey, Jr., W.S. (1976) Organic fluorocompounds in human plasma: prevalence and characterization, in Robert Filler, ed., Biochemistry Involving Carbon-Fluorine Bond, Amer.Chem.Soc.Symp., series 28, 117-134.

Hall, L.L. et al. (1977) Kinetic model of fluoride metabolism in the rabbit. Environ. Res., 13; 285-302.

Hanhijärvi, H.(1975) Inorganic plasma fluoride concentrations and its renal excretion in certain physiological and pathological conditions in man. Fluoride, 8; 198-207.

Hein, J.W. et al. (1956) Distribution in the soft tissue of the rat of radioactive fluoride administered as sodium fluoride. Nature, 178; 1295-1296.

Hellström, I. (1976) Studies on fluoride distribution in infants and small children. Scand. J. Dent. Res., 84; 119-136.

Hodge, H.C. and Smith, F.A. (1965) In Fluorine Chemistry, ed., Simmons, J.H., Academic, New York, Vol. IV, pp. 518, 535.

Hodge, H.C., Smith, F.A. and Gedalia, I. (1970) Excretion of fluorides. In Fluorides and Human Health, WHO Monograph, series 59, Geneva, p. 141.

Husdan, H. et al. (1976) Serum ionic fluoride: normal range and relationship to age and sex. Clin. Chem., 22; 1884-1888.

Inkovaara, J. et al. (1973) Fluoride and Osteoporosis. Brit. Med. J., 1; 613.

Jardillier, J.C. and Desmet, G. (1973) Study of serum fluorine and its combinations by a technique utilizing a specific electrode. Clin. Chim. Acta, 47; 357-363.

Konikoff, B.S. (1974) The bioavailability of fluoride in milk. J.La.Dent.Assoc., Fall, 7-12.

Kuo, H.C. and Stamm, J.W. (1975) The relationship of creatinine clearance to serum fluoride concentration and urinary fluoride excretion in man. Archs. Oral Biol., 20; 235-238.

Mazze, R.I., Shue, G.L. and Jackson, S.H. (1971a) Renal dysfunction associated with methoxyflurane anesthesia. A
randomized, prospective clinical evaluation. J.A.M.A.,
216; 278-288.

Mazze, R.I., Trudell, J.R. and Cousins, M.J. (1971b) Methoxyflurane metabolism and renal dysfunction: clinical correlation in man. Anesthiol., 35; 247-252.

Mazze, R.I., Cousins, M.J. and Kosek, J.C. (1972) Dose-
related methoxyflurane nephrotoxicity in rats: A biochemical and pathological correlation. Anesthiol., 36; 571-
587.

Parkins, F.M. et al. (1974) Relationships of human plasma
fluoride and bone fluoride to age. Calcif. Tiss. Res.,
16; 335-338.

Raisz, L.G. and Taves, D.R. (1967) The effect of fluoride on
parathyroid function and responsiveness in the rat. Calc.
Tiss. Res., 1; 219-228.

Rich, C., Ensinck, J. and Ivanovich, P. (1964) The effects
of sodium fluoride on calcium metabolism of subjects with
metabolic bone diseases. J. Clin. Invest., 43; 545-556.

Shannon, I.L. et al. (1973) Effect of rate of gland function
on parotid saliva fluoride concentration in the human.
Caries Res., 7;1-10.

Shen, Y.W. and Taves, D.R. (1974) Fluoride concentration in
the human placenta and maternal and cord blood. Amer. J.
Obstet. Gynecol., 119; 205-207.

Shupe, J.L. et al. (1963) The effect of fluorine on dairy
cattle. II. Clinical and pathological effects. Am. J.
Vet. Sci., 24; 964-979.

Simpson, W.J. and Tuba, J. (1968) An investigation of fluoride concentration in the milk of nursing mothers. J.
Oral Med., 23; 104-106.

Singer, L. and Armstrong, W.D. (1960) Regulation of human
plasma fluoride concentration. J. Appl. Physiol., 15;
508-510.

Singer, L. and Armstrong, W.D. (1964) Regulation of plasma
fluoride in rats. Proc. Soc. Exp. Biol. Med., 117;686-689.

Singer, L., Armstrong, W.D. and Vogel, J.J. (1965) In: International Association for Dental Research, 43rd Annual
Meeting, Toronto, Abstracts, Chicago. Am. Dent. Assoc.,
p. 46, Abstr. #38.

Singer, L. and Armstrong, W.D. (1973) Determination of fluoride in ultrafiltrates of sera. Biochem. Med., 8; 415-
422.

Singer, L., Ophang, R.H. and Armstrong, W.D. (1976) Influence of dietary restriction on regulation of plasma and soft tissue fluoride contents. Proc.Soc.Exp.Biol.Med., 151; 627-631.

Singla, V.P., Garg, G.L. and Jolly, S.S. (1976) The Kidneys. Fluoride, 9; 33-35.

Smith, F.A., Gardner, D.E. and Hodge, H.C. (1950) Investigations on the metabolism of fluoride. II. Fluoride content of blood and urine as a function of the fluorine in drinking water. J. Dent. Res., 29; 596-600.

Smith, F.A. et al. (1960) The effects of the absorption of fluoride. AMA Arch. Ind. Health, 21; 330-332.

Suttie, J.W. (1968) Effects of dietary fluoride on the pattern of food intake in the rat and the development of a programmed pellet dispenser. J. Nutr., 96; 529-536.

Taves, D.R. (1970) New approach to the treatment of bone disease with fluoride. Fed. Proc., 29; 1185-1187.

Taves, D.R. and Morrison, A.B. (1970) Fluoride metabolism in experimental chronic renal failure. Fed. Proc., 29; 500, Abstr. #1439.

Taves, D.R. and Neuman, W.F. (1964) Factors controlling calcification in vitro: Fluoride and Magnesium. Arch. Biochem. Biophys., 108; 390-397.

Taylor, J.M., Gardner, D.E. and Scott, J.K. (1961) Toxic effects of fluoride on the rat kidney. II. Chronic Effects. Tox. Appl. Pharm., 3; 290-314.

Wallace-Durbin, P. (1954) The metabolism of fluorine in the rat using F^{18} as a tracer. J. Dent. Res., 33; 789-800.

Walzer, M. and Rahill, W.J. (1966) Renal tubular transport of fluoride compared with chloride. Am. J. Physiol., 210; 1290-1292.

Whitford,G.M.and Taves, D.R. (1971) Fluoride induced diuresis. Plasma concentration in the rat. Proc.Soc.Exp.Biol. Med., 137; 458-460.

Whitford,G.M.and Taves, D.R. (1973) Fluoride-induced diuresis: Renal-tissue solute concentrations, functional, hemodynamic and histologic correlates in the rat. Anesthiol., 39; 416-427.

Whitford, G.M., Pashley, D.H. and Stringer, G.I. (1976) Fluoride renal clearance: a pH-dependent event. Am. J. Physiol., 230; 527-532.

Whitford, G.M. and Taves, D.R. (1971) Fluoride induced diuresis. Plasma concentration in the rat. Proc. Soc. Exp. Biol. Med., 137;458-460.

Whitford, G.M. and Taves, D.R. (1973) Fluoride-induced diuresis: Renal-tissue solute concentrations, functional, hemodynamic and histologic correlates in the rat. Anesthiol., 39;416-427.

Whitford, G.M., Pashley, D.H. and Stringer, G.I. (1976) Fluoride renal clearance: a pH-dependent event. Am. J. Physiol., 230;527-532.

Williams Obstetrics (1976), 15th ed., Eds. J.A. Pritchard and P.C. MacDonald, Appleton. Century Crofts, New York, p. 173.

Yao, K. and Grøn, P. (1970) Fluoride concentration in the duct saliva and in whole saliva. Caries Res., 4;321-331.

Zipkin, I. et al. (1956) Urinary fluoride levels associated with use of fluoridated waters. Pub. Hlth. Rep., 71;767-772.

8

The Effect of Body Fluid pH
on Fluoride Distribution,
Toxicity, and Renal Clearance

Gary M. Whitford and David H. Pashley

Renal excretion represents the most important avenue
for the removal of fluoride from the body. Most reports
indicate that 40% to 60% of the fluoride ingested daily is
excreted in the urine. Fluoride enters the tubules of the
kidney, as far as is known, exclusively with the ultra-
filtrate from the glomeruli. The amount of fluoride entering
the nephrons per unit time is determined by two factors, the
glomerular filtration rate and the plasma fluoride concentration. The amount of fluoride excreted in the urine per
unit time, however, has consistently been found to be less
than the amount which enters the kidney tubules. Thus,
fluoride handling by the kidney is characterized by filtration at the glomeruli followed to a varying degree by
tubular reabsorption. The reabsorbed fraction of filtered
fluoride varies from as little as 20% to as much as 95%.
From these extremes, it is apparent that the total body
burden of fluoride is, to a considerable extent, determined
by fluoride renal clearance.

Previous attempts to define the mechanism of the renal
handling of fluoride are in general agreement but did not
take into account all of the data. Chen et al. (1) measured
the renal clearance of fluoride in dogs under a variety of
diuretic states. During mannitol or saline diuresis, it was
noted that fluoride clearance correlated well with urinary
flow rate. It was suggested that fluoride ion, which has a
large hydrated radius, is transported in bulk flow with the
tubular fluid. Similar conclusions were reached by Carlson
et al. (2). Carlson et al. (3) studied the renal clearance
of radiofluoride in man. They manipulated their urinary
flow rates by ingesting water and confirmed the relationship between fluoride clearance and urinary flow rate.
Walser and Rahill (4), comparing the clearances of chloride
and radiofluoride in the dog, concluded that the reab-
sorptions of fluoride and chloride varied together when

chloride clearance was high but that fluoride clearance correlated best with urine flow rate when chloride clearance was low. The main conclusion which has been drawn from the diuresis experiments is that fluoride clearance is positively correlated with urine flow rate. However, during diuresis induced by nitrate or sulfate in dogs, Chen et al. (1) observed that fluoride clearance did not increase with urinary flow rate even though the magnitude of the flow was similar to that achieved during mannitol and saline diureses. No interpretation was offered for this lack of correlation. It suggested to us that fluoride clearance might not be mechanistically dependent on urinary flow rate.

Urinary pH and

the Renal Clearance of Fluoride

Since the suggested mechanism for the renal handling of fluoride was unable to explain several reported observations, a series of experiments were designed to further investigate the mechanism of renal fluoride reabsorption and the relationship of fluoride clearance with urinary flow rate, clearances of the major endogenous solutes (sodium, potassium, and chloride), osmolar clearance, and, in some experiments, excretions of calcium and phosphate.

Studies in the Rat

A variety of protocols were used, including the infusions of hypo-, iso-, and hypertonic saline and mannitol. The results of these experiments generally agreed with those of the earlier investigations. That is, fluoride clearance was often positively correlated with urine flow rate. However, a statistically significant correlation was not always observed and, occasionally, a negative correlation was found. Another problem was that whenever fluoride clearance did correlate with urinary flow rate, it also correlated with practically every other variable measured. Thus, because the factors could not be separated, it was not possible to conclude that the renal handling of fluoride was mechanistically linked to any measured variable.

Whitford et al. (5) found by comparing the effects of two diuretics, acetazolamide (Diamox*) and furosemide (Lasix+), that urine pH might be important in the renal handling of fluoride. These two drugs have different

---

*Lederle Laboratories, Pearl River, N.Y.
+Hoechst Pharmaceuticals, Somerville, N.J.

mechanisms of action. Furosemide is the more potent of the two and would, on an equimolar basis, produce a greater diuresis. It also produces a profound chloruresis principally by inhibiting chloride reabsorption in the ascending limb of the loop of Henle (6). Acetazolamide, on the other hand, inhibits the enzyme carbonic anhydrase. The renal tubular reabsorption of bicarbonate ion is inhibited and the urine becomes alkaline.

Ten rats were anesthetized with pentobarbital. All were infused intravenously with a fluoride-containing iso-tonic saline solution. ^{14}C-inulin was included and its clearance used to estimate glomerular filtration rate (GFR). A 30 minute control clearance period was allowed and then five of the animals were given an iv injection of furosemide while the other five received acetazolamide. Two additional clearance periods then followed.

Figure 1 shows the fractional fluoride excretion (C_F/GFR), in percentage units, as a function of urinary flow rate. The fractional fluoride excretion values represent the percent of the amount of fluoride which entered the kidney tubules by filtration during the clearance period which was excreted. It is important to factor fluoride clearance (C_F) by glomerular filtration rate since changes in the latter will influence the former. For the control data, the values were between 10% and 15% indicating that 85% to 90% of the filtered fluoride was being reabsorbed and returned to the systemic circulation. It is apparent that the excreted fraction of the filtered fluoride increased with urinary flow rate in each group. However, the rate of change appeared to be nearly ten times higher in the acetazolamide (Diamox) group.

Figure 2 shows the relationship between the fractional excretions of fluoride and chloride. The well-known chloruretic effect of furosemide was seen. In the acetazolamide group, there was only a slight increase in chloride excretion (bicarbonate becomes the major excreted anion). The rate of change in fluoride excretion with changes in chloride excretion was 50 times higher in the acetazolamide group. This experiment thus suggested, for the first time, that urine pH might play a major role in the renal handling of fluoride.

The next experiment was designed to more carefully examine this possibility (5). In one group of five rats a metabolic acidosis was induced by preadministration of ammonium chloride. These initially acidotic rats were titrated into alkalosis during the course of the renal

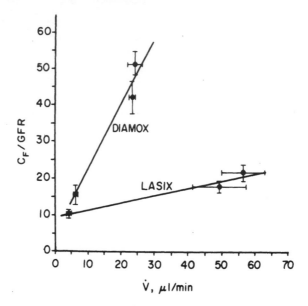

Figure 1. Influence of Diamox (acetazolamide) and Lasix (furosemide) on fractional fluoride excretion as a function of urine flow rate in rats. Left points for each group are from pre-diuretic control clearances. Mean ± SEM.

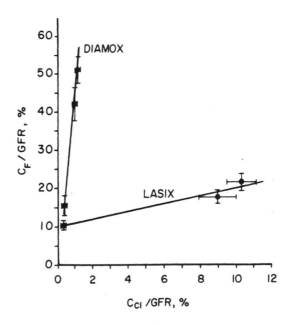

Figure 2. Influence of Diamox (acetazolamide) and Lasix (furosemide) on fractional fluoride excretion as a function of fractional chloride excretion in rats.

clearance study by intravenous infusion of sodium bicarbonate and acetazolamide. A second group of six rats was
initially made alkalotic by infusion of sodium bicarbonate.
The infusate was changed to isotonic saline during the
experiment and the rats were titrated into acidosis by
administration of ammonium chloride. At the mid-point of
each half-hour clearance period, urinary pH was determined
anaerobically.

The relationship between urinary pH and fractional
fluoride excretion is shown in Figure 3. When the pH was
5.6 or less, the excreted fraction of filtered fluoride was
less than five percent, i.e., fluoride reabsorption was in
excess of 95%. Above pH 5.6, there was a progressive increase in fractional fluoride excretion which reached a
maximum mean value of nearly 70%. In both the initially
acidotic and initially alkalotic groups, fractional
fluoride excretion correlated strongly with urine pH
($p < .001$). Both these results and those of the furosemide-
acetazolamide experiment implicate urine pH as an important
determinant of the renal clearance of fluoride.

The relationship between urinary flow rate and pH
during a forced diuresis was also examined (5). Figure 4
summarizes results from six rats undergoing a progressive
mannitol diuresis. During periods one and two, an isotonic
mannitol solution containing ^{14}C-inulin was infused at a low
rate (16.7 µl/min). Urine flow rates were about 10 µl/min.
During periods three and four the infusion rate was doubled
and urine flow rate increased almost proportionally. The
infusion rate was increased again for periods five and six,
and again for periods seven and eight. As expected,
fractional fluoride excretion increased concurrently with
urine flow rate. At the same time, however, urine pH was
also rising. The decrease in hydrogen ion concentration
showed a positive relationship with urine flow rate which
suggested that, during a mannitol diuresis, the rate of
tubular hydrogen ion secretion remains relatively constant
and pH rises due to increased volume flow, that is, by
dilution. Thus, the positive correlation between urinary
flow rate and fluoride clearance may simply reflect an
increasing urinary pH.

The experiment was taken one step further in an attempt
to dissociate fluoride clearance from urinary flow rate.
During period nine the mannitol infusion was replaced with
an isotonic sodium bicarbonate infusate containing acetazolamide. The infusion rate was reduced from 150 to 25

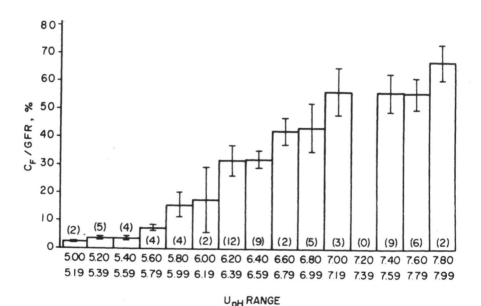

Figure 3. Dependence of fractional fluoride excretion on urine pH in rats. Five rats were initially acidotic and titrated into alkalosis. Six rats were initially alkalotic and titrated into acidosis. Urine pH is shown in 0.2 unit intervals. The number of urine samples falling in a given 0.2 pH unit interval is shown in parentheses. Reproduced with permission of The American Physiological Society.

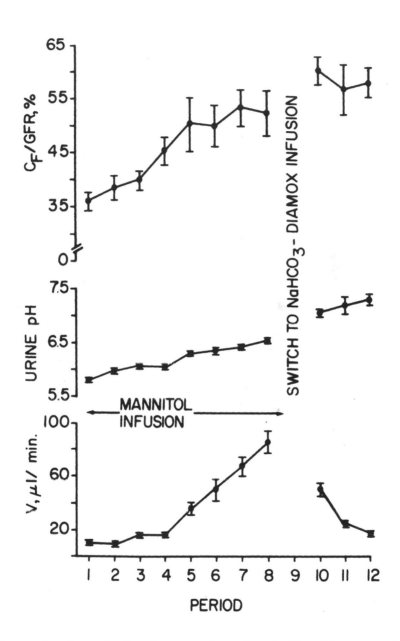

Figure 4. Association between fractional fluoride excretion, urine flow rate, and urine pH during a graded mannitol diuresis in rats (left). Dissociation of fractional fluoride excretion from urine flow rate and its association with urine pH during a declining diuresis and alkalinization of the urine (right). Reproduced with permission of The American Physiological Society.

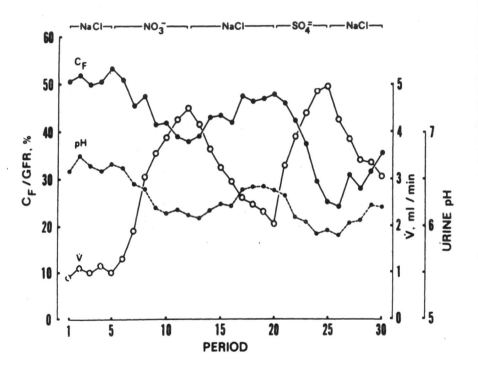

Figure 5. Dissociation of fractional fluoride excretion from urine flow rate and its association with urine pH during nitrate and sulfate diuresis in the dog.

μl/min. This produced a precipitous fall in urine flow rate
(Figure 4) and chloride excretion (not shown). However,
both urine pH and fractional fluoride excretion showed
further increases.

Studies in the Dog

As noted previously, the renal clearance of fluoride
failed to correlate with urine flow rate during nitrate and
sulfate diuresis (1). Rector (7) suggested that these
poorly resorbable anions should enhance hydrogen ion
secretion by creating an increase in the distal transtubular
potential and should result in an acidification of the urine.
Figure 5 shows the results of experiments with a nonanesthe-
tized dog which were designed to evaluate the changes in
urine flow rate, urine pH, and fractional fluoride excretion
during the infusion of nitrate or sulfate (8).

During the one-hour equilibration period and the first
five 20 minute clearance periods, the animal received an
isotonic saline infusion (2 ml/min) which also delivered
inulin and fluoride. Urine flow rate was approximately
1 ml/min, urine pH was 6.6 to 6.8, and fractional fluoride
excretion was 50% during this time. The saline infusate
was then replaced with one containing isotonic sodium
nitrate and the infusion rate was doubled to 4 ml/min for
the next seven periods. Osmotic diuresis was promptly
established and peaked at 4.5 ml/min. Both urine pH and
fractional fluoride excretion, however, decreased moderately.
The original saline infusion was then restarted and continued for eight more clearance periods. During this saline
recovery period, urine flow rate decreased progressively
to a low of 2 ml/min while urine pH and fractional fluoride
excretion trended upward. The recovery period was followed
by five periods of isotonic sodium sulfate infusion and
then a final saline recovery period. The results during
these phases were qualitatively similar to those observed
during nitrate infusion and the consecutive recovery period.
This experiment provides an explanation for the findings of
Chen et al (1) and gives further evidence that the renal
clearance of fluoride is not dependent on urine flow rate.
Thus, several techniques exist for dissociating fluoride
clearance from urine flow rate. On the other hand, attempts
to dissociate fluoride renal clearance from urine pH have
not been successful.

Studies in Man

We have obtained preliminary data (unpublished) on the
influence of acute metabolic alkalosis on fluoride excretion

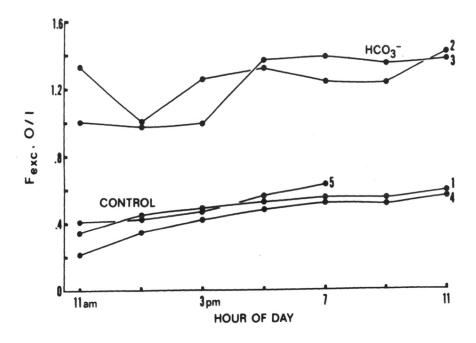

Figure 6. Fluoride urinary excretion as a fraction of fluoride intake in liquids (0/I) during normal acid-base balance (Days 1, 4, 5) and alkalosis (Days 2, 3) by a human.

in human volunteers. Figure 6 shows the fraction of fluoride ingested in liquids (1.6-2.2 mg/day) which was excreted in the urine as a function of time of day. The data is from a single volunteer over five consecutive days. During the first day the subject's excretion accounted for 35% to 60% of the ingested fluoride. Urinary pH was determined anaerobically at each 2-hour voiding and ranged from 5.8 to 6.7. During days two and three, the subject ingested 1-2 g sodium bicarbonate every two hours and the resulting urine pH ranged from 6.8 to 7.5. In this period of metabolic alkalosis, the excreted amount of fluoride ranged from 1 to 1.4 times the amount accounted for by fluid intake. The excess must have been derived from solid food or from mobilization of body fluoride stores. The subject then stopped the sodium bicarbonate regimen and continued urine collections for two more days. Urinary pH declined accordingly, as did the fractional fluoride excretion.

In another experiment four normal human volunteers ingested 100 ml of water and voided hourly throughout each experimental day. The 7-hour time course for excretion of 10 mg, or 520 μmole, doses of fluoride (as NaF) under normal acid-base conditions (urinary pH 5.8-6.6) and during a bicarbonate-induced alkalosis (urinary pH 6.5-7.4) is shown in Figure 7. During the first four hours under control conditions, the fluoride excretion rate averaged 24.6 μmole/hour, and it was twice as high (48.3 μmole/hour) during alkalosis. Over the 7-hour observation periods, 60% more fluoride was excreted under alkalosis as compared to normal (control) conditions.

These preliminary results are in accord with the data from the rat and dog studies and suggest that the renal clearance of fluoride in the human is also a pH-dependent process.

Proposed Mechanism for Renal Reabsorption of Fluoride

To account for the pH-dependence of the renal transport of fluoride, we have hypothesized that the permeating moiety is hydrogen fluoride. The transport process is outlined in Figure 8.

The fluoride ion (F^-), because of its charge and large hydrated radius, is considered to have poor ability to permeate the tubular epithelium. The undissociated HF molecule is considered very permeable. In an alkaline urine, the relative concentration of F^- will be high. For example, at a pH of 8, the F^- to HF concentration ratio is 3.5×10^4 according to the Henderson-Hasselbach equation. The poorly

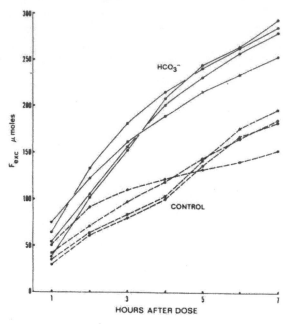

Figure 7. Seven hour time course of urinary excretion of ingested fluoride (520 μmoles or 10 mg as NaF) by four human volunteers during normal acid-base balance and alkalosis.

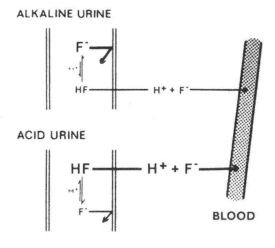

Figure 8. Proposed mechanism of fluoride renal reabsorption. Hydrogen fluoride is regarded as the permeating moiety. The process occurs by hydrogen fluoride diffusion from the tubule and F^- trapping in the interstitium.

permeating fluoride ion remains within the tubule to be
excreted and fluoride clearance is high. Some reabsorption
occurs, however, even when the urinary pH is near 8. HF may
still be the resorbable form of fluoride, however, because
even though the final urinary pH is high, micropuncture
studies have shown that the intratubular pH through the
proximal tubule and most of the distal tubule is less than
that of blood under all acid-base conditions (9). Upon
entering the interstitial fluid, where the pH is high and
the fluoride concentration low (relative to the tubular
concentration), hydrogen fluoride must dissociate and
release fluoride ion. The ion then diffuses into the
"leaky" peritubular capillaries and is returned to the systemic circulation. In a more acid intratubular fluid, the
transepithelial hydrogen fluoride concentration gradient is
increased. The reabsorptive process proceeds according to
the same mechanism as described above but at a higher rate.
Thus, less fluoride remains for excretion.

The overall reabsorptive process occurs by nonionic
diffusion of hydrogen fluoride from the tubule and trapping
of fluoride ion in the interstitium. Over the pH range of
intratubular fluid (4.5-8.0), the largest fraction of
fluoride exists in the ionic form which serves as a reservoir for hydrogen fluoride. The activity ratio of fluoride
ion to hydrogen fluoride must remain constant at any given
pH as defined by the Henderson-Hasselbach equation. Hence,
as hydrogen fluoride diffuses from the tubule, tubular ionic
fluoride will combine with hydrogen ions to maintain
equilibrium conditions. The continuous removal of fluoride
from the interstitium by the peritubular blood flow and of
water from the tubular lumen by normal renal mechanisms
assures that the diffusion gradient for hydrogen fluoride is
directed from the tubule to the interstitium. The process
is analogous to, but in the reverse direction of, ammonia
secretion which occurs by nonionic diffusion and the
subsequent trapping of ammonium ion within the tubule (10).

Fluoride Absorption

from the Urinary Bladder

To further evaluate the above fluoride transport
hypothesis, Whitford et al. (11) studied the systemic absorption of radioactive (^{18}F) and stable fluoride from the
urinary bladder of the rat. This model was selected because
the mammalian urinary bladder has a very "tight" epithelium
which severely restricts the permeation of most solutes
(10,12). Several preliminary experiments were performed to
characterize the system, including an assessment of the

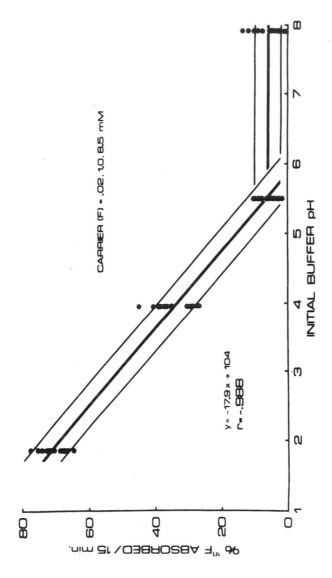

Figure 9. pH-dependence of ^{18}F absorption, as percent of initial dose, from urinary bladder over pH range 1.85-5.50. Equation refers to this portion of graph. Data from individual rats in each carrier-fluoride concentration group are plotted. pH-independent absorption occurs above pH 5.50. Mean and standard errors of two functions are shown.

influence of intravesicular volume and pressure, the rate of
^{14}C-inulin absorption, the influence of buffer selection,
and the influence of urinary constituents. The volume used,
200 μl, was within the range of urinary volumes commonly
removed from the bladders of anesthetized rats; inulin absorption was about one percent per hour; and the buffer
selection (except as it influenced pH) and the presence or
absence of urine did not significantly influence the
fluoride absorption results.

The main experiment was conducted as follows. Sixteen
rats were randomly assigned to four groups based on the
solution pH. The pH values were approximately 2, 4, 5.5,
and 8. All solutions contained ^{18}F, were isotonic and, in
the first experiment, the stable fluoride concentration
was 0.02 mM. This concentration is one fifth to one tenth
of that found in normal rat urine. The ureters were ligated
via a mid-line incision and the incision was closed. Test
solutions were introduced through indwelling catheters into
the bladders. After 15 minutes, the solutions were removed
and analyzed for stable and radiofluoride, and ^{14}C-inulin.
Changes in inulin concentrations were minor but were,
nevertheless, used to correct the fluoride data for water
movement. This basic experiment was repeated with stable
fluoride concentrations of 1.0 and 8.5 mM.

Figure 9 shows the overall results in terms of the percentage of fluoride absorbed in 15 minutes at the various
initial buffer pH values. Two important conclusions can be
drawn from these results. (1) Up to a pH value of approximately 6, fluoride absorption is inversely related to pH.
That is, fluoride absorption is greatest when the hydrogen
ion concentration (and, therefore, the hydrogen fluoride
concentration) is highest. (2) The fractional absorption
per unit time is not influenced by the initial fluoride
concentration. For example, from a buffer solution of pH 2,
70% of the fluoride was absorbed in 15 minutes whether the
initial fluoride concentration was 0.02, 1.0, or 8.5 mM.
Thus, the fractional absorption is independent of concentration, a characteristic of a first-order process such as
passive diffusion. Both of these observations support the
hypothesis that fluoride transport occurs by the nonionic
diffusion of hydrogen fluoride.

A further finding of considerable interest concerns
the relative permeabilities to fluoride of the renal tubular
and urinary bladder epithelia. Permeability coefficients
for these two epithelia, estimated on the basis of published
surface area data for the tubules (13) and directly measured
area for the bladder (11), were virtually identical for pH

values in the mid to low 5 range (3.3×10^{-6} cm sec^{-1}). This agreement is remarkable considering the structures and the physiologic roles of the two epithelia. These data suggest that hydrogen fluoride permeates the urinary bladder more easily than most compounds yet studied ([14],[15],[16],[17],[18],[19], [20]).

Fluoride Distribution and Excretion

in Chronic Acid-Base Disturbances

The influence of chronic acid-base disturbances on fluoride metabolism has been studied in rats (in preparation). Twenty-four female F-344 rats (body weight 160 gm) were assigned to three groups and caged in pairs. During Phase 1 of the experiment, the rats had free access to distilled water and standard laboratory rat pellets for two weeks. During Phase 2 (one week) and Phase 3 (three weeks), mild acid-base disturbances were induced. One group received 0.25 M ammonium chloride and 1% mannitol in its fluoride-free drinking water, another received 1% mannitol only, and the third group received 0.15 M sodium bicarbonate. During the final three weeks (Phase 3) the drinking water was unchanged but each rat received a subcutaneous fluoride injection of 6.5 μmol twice daily (approximately 0.75 mg/kg per injection as NaF). Injection was used to insure that all animals received equal amounts of supplemental fluoride. Three 24-hour urine collections were made in each of the first two phases and seven such collections were made during the final two weeks of Phase 3. Body weight, food intake, fluoride-free water intake, and fecal excretion were monitored throughout the study. The only variable which showed significant differences between the groups was water intake which was moderately elevated during alkalosis. On noncollection days, spontaneously voided urine samples were collected directly from the urethral orifice and kept under oil for the determination of pH and pCO_2. On the final day of the experiment, one half of the rats (four from each of the three groups) were sacrificed at two hours and half at four hours after injection of the last subcutaneous dose which contained ^{18}F. A terminal arterial blood sample and a variety of tissue samples were taken. These were analyzed for stable and radioactive fluoride as well as for the major endogenous ions.

The acid-base disturbances were slight. For example, the mean terminal blood pH values were 7.41 and 7.53, and bicarbonate concentrations were 15.2 and 25.4 meq/L, for the acidotic and alkalotic groups, respectively. Mean urine pH values for the acidotic and alkalotic groups during Phase 3

were 5.38 and 6.51, respectively. Values from the control
group were intermediate. Plasma sodium, potassium, chloride,
and calcium concentrations were within normal limits and did
not differ among the groups.

Figure 10 shows the results from the 24-hour urine collections.   The left points for each group are from
Phase 1 when all rats were drinking distilled water. There
were no differences among the groups in urine output, or in
chloride or fluoride excretion. During Phase 2, when the
rats were entering their acid-base disturbances, the acidotic group's fluoride excretion fell from 1.5 to 1.0 μmol/Day.
Chloride excretion increased sharply, due to the ingestion
of ammonium chloride, while urine output was not changed.
Fluoride excretions in the control and alkalotic groups
increased moderately and were significantly higher than in
the acidotic group ($p < .01$). Neither urine flow rate nor
chloride excretion in the control or alkalotic groups were
significantly different as compared to Phase 1. Finally,
during fluoride supplementation in Phase 3, fluoride excretion in the control group was 65% higher than in the
acidotic group and it was twice as high in the alkalotic
group. The renal clearance of fluoride in the alkalotic
group was about four times higher than in the acidotic
group since, as shown in Figure 11, the plasma fluoride
concentration of the acidotic group was nearly twice as high
as that of the alkalotic group.

Figure 11 shows the terminal radiofluoride concentrations, in thousands of counts per minute/ml of tissue water,
in plasma and soft tissues obtained two hours after the
final dose. From left to right for each tissue, the points
are from the acidotic, the control, and the alkalotic group.
For example, the plasma concentrations were 11.9, 8.7, and
6.5 thousand counts per minute/ml respectively, for these
groups. It is apparent that plasma fluoride level is
reflected in each of the eight soft tissues analyzed. That
is, the acidotic group always had the highest concentration
and the alkalotic group always had the lowest concentration.
The whole kidney concentrations are ten times higher than
shown on the graph. Stable fluoride concentrations, both at
two and four hours after the final dose, closely paralleled
these radiofluoride concentrations.

The two-hour ^{18}F concentrations (Figure 11) were
unexpectedly high for brain tissue. Our previous studies
showed values of less than 0.10 for brain tissue-to-plasma
(T/P) ^{18}F concentrations (21).   Carlson et al. (22)
similarly reported a low value in their 80-minute ^{18}F study.
The brain T/P ^{18}F values in the current study were 0.37 ±

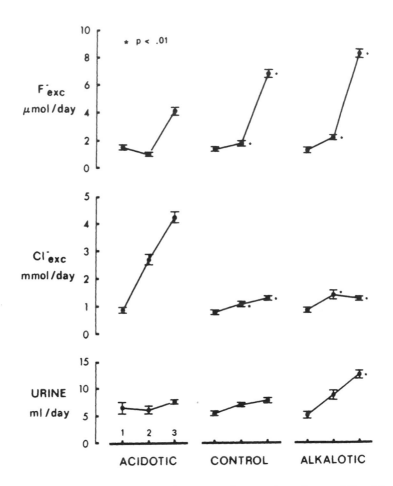

Figure 10.   Daily (24-hour) excretions of urine, chloride and fluoride during chronic acid-base disturbances. Left-to-right for each group: Phase 1 - All rats on distilled water; Phase 2 - Acidotic group, 0.25 M ammonium chloride in drinking water; control group, distilled water; alkalotic group, 0.15 M sodium bicarbonate. Phase 3 - Same as Phase 2 but each rat received 6.5 µmol fluoride subcutaneously twice daily. Significance testing (p < .05 by t-test, asterisks) compares control and alkalotic groups to the acidotic group in the same phase.

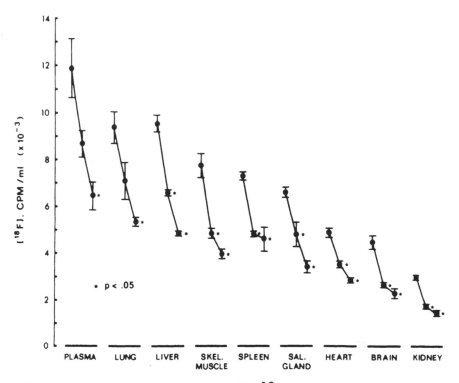

Figure 11. Tissue water fluoride (^{18}F) concentrations (mean ± SEM, n = 4) two hours after final subcutaneous injections in the chronic acid-base disturbance study. Kidney concentrations are 1 x 10^{-4}. Left-to-right for each tissue: acidotic, control and alkalotic groups. Significance testing (p < .05 by t-test, asterisks) compares control and alkalotic groups to the acidotic group.

.08, 0.27 ± .03, and 0.35 ± .03 (mean ± SEM throughout) for the acidotic, the control, and the alkalotic group, respectively. Brain tissue-to-plasma values for stable fluoride were 70% to 130% higher than the ^{18}F values both at two and four hours after the final dose. These elevated tissue-to-plasma values were not related to acid-base status. They may reflect an altered permeability of the blood-brain barrier to fluoride or locally high fluoride pools induced by the chronic fluoride regimen.

This experiment clearly demonstrates that mild, chronic acid-base disturbances can markedly influence the retention and soft tissue concentrations of fluoride. This result is due, in part, to the differences in the renal clearance of fluoride in the various acid-base states. The calcified tissue compartment may also be involved in a complex manner. For example, the elevated plasma levels associated with acidosis would increase the amount of fluoride taken up by bone. The calcium loss (Table 1) would indicate excess bone breakdown with the release of fluoride into the plasma.

Bone resorption rate is increased in acidosis (23). This is in keeping with the urinary Ca^{++} excretion data obtained during the experiment which is shown in Table 1.

Table 1. 24-Hour Urinary Calcium Excretions (μmol/day) During Chronic Acid-Base Disturbances.

	GROUP		
	Acidotic	Control	Alkalotic
Phase 1	37.7 ±6.8 (11)	28.1 ±3.7 (12)	28.1 ±4.2 (12)
Phase 2	61.9* ±8.8 (12)	34.3 ±4.2 (12)	30.1 ±3.7 (12)
Phase 3	101.5* ±8.3 (26)	30.1 ±2.4 (24)	14.2* ±1.5 (27)

$\bar{x}$ ± SEM (n). *$p < .05$ (compared to previous phase).

The acidotic group increased its Ca^{++} excretion from about 38 to 62 μmol/day between Phase 1 and Phase 2. During Phase 3, the rate was still higher (over 100 μmol/day). These findings are in agreement with the results of Albright and Reifenstein (24). Glassman et al. (25) further discussed this point and reported enhanced fractional excretions of calcium (C_{Ca}/GFR) and phosphorus during an acute metabolic acidosis in dogs. The control group's Ca^{++} excretion (Table 1) was virtually constant throughout the experiment at approximately 30 µmol/day. Alkalosis, at least during its onset in Phase two did not seem to alter Ca^{++} excretion. However, during Phase three (fluoride supplementation) Ca^{++} excretion fell sharply to 14 µmol/day. This may reflect the stabilizing influence of fluoride ([26,27,28]) and/or alkalosis ([23]) on bone mineral, although no such influence attributable to fluoride was apparent in the control or acidotic groups.

In view of the Ca++ excretion data and their suggestions concerning calcified tissue metabolism, one might have expected fluoride uptake by the bones and teeth of the acidotic animals to have been lowest. In fact, the opposite was true for radiofluorides in femur and incisor as shown in Table 2.

Table 2. Two-Hour Radiofluoride Concentrations in Femur and Incisor as a Function of Acid-Base Status.

	GROUP		
	Acidotic	Control	Alkalotic
Femur	480615 ±34844	394684 ±16822	375219* ±24119
Incisor	860367 ±20745	727851* ±21775	664301* ±50100

CPM/gm wet weight. $\bar{x}$ ± SEM, n = 4.

*p < .05 (compared to Acidotic Group).

For both of these tissues, the acidotic group had the highest ^{18}F concentrations and the alkalotic group had the lowest. These concentration differences probably reflect the different plasma concentrations of the three groups. The uptake of ^{18}F by long bones appears to occur mainly at the epiphyses and periosteal regions by iso- and heteroionic exchange ([29]). The tissue-to-plasma ^{18}F concentration ratios for femur and incisor tended to be lowest in the acidotic group and highest in the alkalotic group although statistical significance was reached only when the incisor

data of the acidotic and alkalotic groups were compared.
Thus, the calcified tissue compartments may have contributed
to the differences in the plasma and soft tissue fluoride
concentrations observed among the groups. Stable fluoride
analyses in the femurs and developing enamel in the incisors
are still in progress and will be reported elsewhere.

Acute Fluoride Toxicity

and Acid-Base Status

It was expected that animals with a pre-existing
alkalosis should be more resistant to acute fluoride toxicity
than animals in acidosis. In alkalosis, the renal clearance
of fluoride is greater and body fluid fluoride concentrations
should, therefore, be lower as compared with acidosis.
Furthermore, it is of interest to consider the possibility
that the soft tissue distribution of fluoride is determined
by the diffusion equilibrium of hydrogen fluoride. If
intracellular pH values are relatively independent of extracellular pH, as reported for skeletal muscle (30,31), heart
(32), and leucocytes (33), then at any given extracellular
fluoride concentration, intracellular fluoride concentrations
should be lower in the alkalotic animal as compared with
normal or acidotic animals. The ratio (R) of the intracellular to extracellular fluoride concentrations can be
calculated from a modification of the Henderson-Hasselbach
equation as follows:

$$R = \{1 + 10^{(pH_i - pK_a)}\}/\{1 + 10^{(pH_e - pK_a)}\} \qquad (1)$$

where pH_i = intracellular pH; pH_e = extracellular pH; and
pK_a = 3.45. If the pH_e were 7.4 and the pH_i of a given
tissue were 6.9, then R = 0.32. If an acidosis were imposed
such that pH_e fell to 7.0 while pH_i remained unchanged at
6.9, then R would equal 0.79; i.e., the intracellular
fluoride concentration would be nearly 2.5 times higher in
acidosis than in alkalosis at the same extracellular
fluoride concentration.

The influence of acid-base status on acute fluoride
toxicity was studied in female F-344 rats (Reynolds et al.,
Tox. Appl. Pharmacol., in press). Body weights were
171 ± 3 gm. Six animals were acidotic following intra-
gastric administration of 3 mmol ammonium chloride at 24
hours and again at 3 hours prior to anesthesia (40 mg/kg
pentobarbital, ip). Eight animals were alkalotic following
the administration of the same quantity of sodium bicarbonate. Continuous intravenous infusions were delivered at

25 µl/min and urine was collected each half hour from indwelling bladder catheters. Arterial blood pressures were monitored and blood samples were collected anaerobically from carotid artery cannulae at the mid-point of each urine collection period. Glomerular filtration rate was determined from the clearance of ^{14}C-inulin. During a one-hour inulin equilibration period and the first half-hour clearance period, each animal received fluoride at a rate of 159 nmol/min. The fluoride infusion rate was then increased to 1340 nmol/min and the infusion was continued until the death of the animal. Urine and blood pH and PCO_2 were determined anaerobically on all rats during the first clearance period (low fluoride) and on selected samples during subsequent periods.

The initial mean blood pH, PCO_2, and bicarbonate concentrations were 7.05, 36.6 mm Hg, 10.0 mM, and 7.42, 44.2 mm Hg, and 27.5 mM in the acidotic and the alkalotic group, respectively. Table 3 contains other pertinent data from this study. Data from the second period are not shown because the rapidly rising plasma fluoride concentrations precluded an accurate estimate of the mean concentrations. Plasma fluoride concentrations ($[F]_p$) were consistently higher in the acidotic group, and this difference was statistically significant in clearance periods five, six and eight. This difference undoubtedly reflects the higher renal clearance of fluoride (C_F) in the alkalotic group ($p < .01$ in every period except seven) which, in turn, can be interpreted in terms of the higher urine pH of this group (5), at least early in the experiment. There was a tendency for the urine pH to rise with time in the acidotic group and to fall in the alkalotic group. The explanation for this is not evident since, as the study progressed, the blood pH declined slightly in the acidotic group while it rose slightly in the alkalotic group. Linear regression analysis of renal fluoride clearance as a function of urine pH showed a statistically significant relationship ($p < .001$). A similarly strong correlation was noted between renal fluoride clearance and glomerular filtration rate (GFR). The progressive decline in glomerular filtration rate was accompanied by a progressive decline in mean arterial blood pressure (MABP).

The acidotic animals began to expire in period seven and were all dead in period nine. Whenever possible, a plasma sample was taken just before death. (The best indicator of impending death was declining blood pressure). The acidotic group's mean terminal plasma fluoride concentration was 1.79 ± .09 mM (n = 5). One alkalotic rat died during period 11, three died during period 13,

Table 3. Time Courses of Plasma Fluoride Concentration, Fluoride Renal Clearance, Glomerular Filtration Rate, Urine pH and Mean Arterial Blood Pressure in Alkalotic and Acidotic Rats Receiving Infusions of Lethal Amounts of Fluoride.

Period	$[F]_p$, mM/L		C_F, ml/min		GFR, ml/min		Urine pH		MAPB, mm Hg	
	A	B	A	B	A	B	A	B	A	B
1	.095	.088	.227	.666*	1.103	1.134	5.34	7.38	125	129
	±.011	±.006	±.043	±.104	±.079	±.210	--	--	±4	±3
	6	6	6	6	6	6	6	8	6	8
3	.682	.557	.265	.465*	.835	1.119	5.78	7.05	113	118
	±.062	±.060	±.034	±.051	±.066	±.154	--	--	±6	±8
	6	8	6	8	6	8	6	2	6	8
4	.888	.647	.208	.371*	.594	.836	6.00	6.30	94	107
	±.107	±.066	±.048	±.038	±.093	±.105	--	--	±10	±10
	6	7	6	7	6	8	3	3	6	8
5	1.107	.754*	.084	.434*	.281	.913*	--	--	84	105
	±.141	±.066	±.035	±.049	±.112	±.076	--	--	±10	±7
	6	7	5	7	5	8	--	--	6	8
6	1.419	.852*	.059	.339*	.365	.753	--	6.58	56	90*
	±.194	±.072	±.022	±.057	±.151	±.150	--	--	±9	±6
	6	7	3	7	2	7	--	5	6	8

	C1	C2	C3	C4	C5	C6	C7	C8	C9	C10
7	1.270 ±.138 4	.999 ±.111 7	.041 ±.036 2	.240 ±.087 7	.271 ±.242 2	.512 ±.121 7	--- --- ---	--- 6.39 2	47 ±13 4	84* ±9 8
8	1.691 ±.102 2	1.212* ±.128 7	--- --- ---	.138 ±.060 7	--- --- ---	.445 ±.113 7	--- --- ---	--- 6.46 3	48 ±0 2	70* ±6 8
9	--- --- ---	1.456 ±.191 6	--- --- ---	.129 ±.068 6	--- --- ---	.363 ±.130 6	--- --- ---	--- --- ---	--- --- ---	70 ±6 5
10	--- --- ---	1.676 ±.250 6	--- --- ---	.086 ±.059 6	--- --- ---	.263 ±.094 6	--- --- ---	--- 6.72 3	--- --- ---	69 ±7 8
11	--- --- ---	1.869 ±.276 6	--- --- ---	.170 ±.130 4	--- --- ---	.235 ±.122 4	--- --- ---	--- --- ---	--- --- ---	57 ±7 8
12	--- --- ---	1.939 ±.242 6	--- --- ---	.103 ±.089 4	--- --- ---	--- --- ---	--- --- ---	--- --- ---	--- --- ---	55 ±5 7
13	--- --- ---	2.127 ±.170 3	--- --- ---	--- --- ---	--- --- ---	--- --- ---	--- --- ---	--- --- ---	--- --- ---	50 ±11 5

Data expressed as mean ± SEM, n. Asterisk indicates $p < .05$.

and the remaining four animals died during period 14. The
terminal plasma fluoride concentration of this group was
2.60 ± .27 mM (n = 6), a value significantly higher than
that of the acidotic group (p < .025). Overall, the mean
acute lethal doses, i.e., the total doses administered, were
29.2 ± 1.7 and 54.7 ± 2.6 mg F/kg for the acidotic and alkalotic groups, respectively (p < .001).

The hearts were removed from three animals in each
group and analyzed for fluoride. The tissue water-to-plasma
fluoride concentrations were 0.63 ± .03 and 0.53 ± .02
(mean ± SEM) in the acidotic and alkalotic groups,
respectively (p < .05). Similar analyses on samples of
abdominal wall skeletal muscle did not reveal a statistically significant difference although the mean value was 40%
lower in the alkalotic animals (0.43 ± .09 vs. 0.28 ± .03).

These data indicate that animals with a pre-existing
alkalosis are less susceptible to the toxic effects of
acutely administered fluoride than are those with a pre-
existing acidosis. A part of the protective effect of
alkalosis is due to an enhanced renal excretion of fluoride
and this result appears to be due to both the influence of
urine pH on fluoride clearance and to a less readily depressed
glomerular filtration rate. However, the alkalotic animals
at death had a mean plasma fluoride concentration which
exceeded that of the acidotic animals by 45%. The explanation for the apparent increase in tolerance afforded by alkalosis may involve lower intracellular fluoride concentrations in critical tissues as suggested by the heart and
skeletal muscle T/P fluoride concentration ratios. Pre-
terminal changes in tissue perfusion may, in part, explain
the different T/P fluoride ratios of the acidotic and
alkalotic groups in skeletal muscle. Such effects, however,
would be less likely in the heart. Further studies are
required to clarify these relationships and these must
include determinations of regional blood flow rates and
intracellular-to-extracellular pH gradients.

Discussion

The results presented show that urinary pH has a
pronounced effect on the renal handling of fluoride and on
fluoride toxicity. The mechanism whereby urinary pH
influences the renal tubular transport of fluoride appears
to involve pH-dependent variations in the concentration
gradient (tubular lumen to interstitium) of undissociated
hydrogen fluoride. It is hypothesized that the tubular
epithelium is not easily permeated by ionic fluoride and
that fluoride transport occurs by the diffusion of HF with

subsequent dissociation and interstitial trapping of ionic fluoride. Further evidence supporting the pH-dependence of fluoride (HF) transport was derived from rat urinary bladder absorption studies. The rate of absorption was a first order process (diffusion) and was dependent on the concentration of HF, i.e., on pH and fluoride concentration.

In chronic, mild acidosis, the urinary excretion of fluoride by rats was reduced and fluoride concentrations in both soft and hard tissues were increased relative to the control and alkalotic groups. Urinary fluoride excretion in the acidotic group was only one half, and renal fluoride clearance was only one quarter, that of the alkalotic group. These findings suggest that the renal mechanism plays an important role in determining body fluoride content during relatively minor disturbances in acid-base balance. An additional finding from this study involved unexpectedly high T/P fluoride ratios for brain. These were unrelated to the acid-base status of the animal and suggested that exposure to elevated fluoride concentrations on a chronic basis alters the fluoride distribution characteristics of this tissue.

A significant protective effect in acute fluoride poisoning was afforded by alkalosis compared to acidosis in rats. The alkalotic animals excreted more fluoride, had a lower plasma fluoride concentration at any given administered dose, could tolerate a dose of nearly 90% more fluoride, and died at a plasma fluoride concentration which was 45% higher than that of the acidotic animals. Tissue-to-plasma fluoride concentration ratios in heart and skeletal muscle suggested that the intracellular concentrations were lower in the alkalotic group at any given plasma fluoride level.

During the past half century, research has consistently indicated that the biologic effects of fluoride are increased as solution pH is decreased. In an extensive review, Borei (34) reported that fluoride inhibition of metabolism in mammalian, plant, and yeast cells was a function of pH. He summarized the work of others who concluded that the pH-dependence of "fluoride inhibition might be explained provided that fluoride penetrates into the cell only as the undissociated HF." In addition to the permeability aspect, hydrogen fluoride may exert a direct metabolic inhibitory effect since the pH-dependence was also evident in broken-cell preparations (34). This possibility is supported by the findings of Lent et al. (35) in their studies on fluoride binding to cytochrome C peroxidase.

Traditionally, however, in vivo studies on fluoride

metabolism in mammals have not been designed or interpreted
in terms of pH, or hydrogen fluoride, gradients. This is
not surprising since, at the pH of plasma and most extracellular fluids, the HF/F$^-$ ratio is about 1/10,000. Thus,
if the plasma fluoride concentration were 1 µM, the concentration of hydrogen fluoride would be vanishingly small
at 0.1 nM. Nevertheless, pH gradients appear to exist
across all adjacent body fluid compartments (36) and,
therefore, the possibility for hydrogen fluoride gradients
must also exist. Even at the normal pH of plasma, the
minute hydrogen fluoride fraction as a determinant of the
overall distribution of fluoride cannot be ignored if the
permeability coefficient of the extensively hydrated
fluoride ion is very low relative to that of undissociated
hydrogen fluoride. These permeability coefficients have
not been determined.

The importance of pH, or hydrogen fluoride, gradients
in the tissue distribution of fluoride can be assessed by
using the DMO technique of Waddell and Butler (37) and
Irvine et al. (38). DMO, a metabolic product of trimetha-
dione, is a weak acid whose pK$_a$ is approximately 6.13 at the
ionic strength and temperature of mammalian extracellular
fluid. The technique independently estimates intracellular
pH. The principal assumption in the use of DMO as an intracellular pH marker is that the ratio of the permeability
coefficients of the nonionized to the ionized fraction
(K_U/K_I) is so large that the diffusion of the ionized species
can be ignored. Boron and Roos (39) confirmed the validity
of the technique by comparing DMO-derived results with those
obtained with glass microelectrodes in giant barnacle muscle
fibers and squid giant axons.

Although the permeability coefficients for hydrogen
fluoride and ionic fluoride are not known, the equations of
the DMO technique (38) can be used to estimate intracellular
pH on the basis of the distribution of fluoride. The
fluoride-derived estimates can then be compared with
published results obtained from the distribution of DMO or
other techniques. The intracellular pH of skeletal muscle
is usually reported as 6.8-7.1 (36). Using T/P fluoride
concentration data for skeletal muscle (21) and assuming a
pK$_a$ of 3.45 and an extracellular pH of 7.4, the intracellular pH is estimated at 7.05, 7.08 and 7.11 for the low,
intermediate and high fluoride groups, respectively. Miyao
(40), using DMO, reported rat liver intracellular pH to be
7.10 and from our fluoride data it is 6.97, 6.96 and 7.25
in the low, intermediate, and high fluoride groups, respectively. Schloerb and Grantham (41) reported the intracellular pH of dog heart to be 6.86. Our rat heart T/P fluoride

data predict intracellular pH values of 6.76 and 6.90 for the intermediate and high fluoride groups, respectively. The intracellular pH of erythrocytes of several species has been reported to range from 7.11-7.27 (36) as determined by electrode and DMO distribution. From the fluoride distribution data in dog erythrocytes (control group) given by Carlson et al. (42), the intracellular pH was estimated at 7.25. The similarities of the fluoride and DMO intracellular pH estimates indicate that the hydrogen fluoride may be the species involved in fluoride diffusion across cell membranes. This hypothesis is supported by our renal clearance and urinary bladder absorption data which show a marked pH-dependence suggesting that K_U/K_I for fluoride is indeed large.

There are, however, some data reported in the literature which appear inconsistent with this hypothesis. In the ^{18}F data of Wallace-Durbin (43) the T/P values for several of the tissues were quite variable over time and were often well in excess of unity. This observation may suggest binding, but the explanation is not clear. Carlson et al. (22) did not find T/P values greater than 1.0, except for tendon which has a high calcium content, nor did we in numerous rat studies with both stable and radiofluoride, except for submandibular salivary glands from rats with no unusual previous exposure to fluoride.

Concluding Remarks

The dependence of fluoride transport on pH gradients has important implications. In cases of acute fluoride poisoning, expeditious manipulation of urinary pH might favorably affect the clinical outcome. Currently, it is recommended that a diuresis be induced in acute toxicity. The results of our studies suggest that such therapy is rational only if there is a concurrent increase in urine pH. Therefore, diuresis should be induced with a urinary alkalinizing agent. Metabolic disorders and certain other disease states which produce acid-base disturbances, such as chronic renal failure, renal tubular acidosis, chronic obstructive pulmonary diseases, and diabetes mellitus, may alter the retention and tissue concentrations of fluoride.

In view of the pH-dependence of renal fluoride transport and the ease with which hydrogen fluoride crosses the generally restrictive bladder epithelium, it may be that the existence of pH gradients across various adjacent body fluid compartments is involved in several aspects of fluoride metabolism. Absorption from the gastrointestinal tract is an obvious possibility. The intracellular-to-extracellular

partition of fluoride may also be related to pH, or hydrogen
fluoride, gradients.

The altered distribution of fluoride in brain when the
fluoride is given on a chronic basis is interesting in light
of the functional changes in rats and Rhesus monkeys reported
by Lu et al. (44,45). These authors noted an increased
central nervous system sensitivity to chemical and physical
stimuli in both species at relatively low, but chronic,
fluoride doses (7 and 70 ppm dietary fluoride for several
weeks; 2 mg/kg daily for eight weeks in monkeys).

In caries prevention, it may be desirable to recommend
the ingestion of supplemental fluoride just before bedtime
rather than at breakfast as is usually done. During sleep,
voiding is less frequent or absent and both urine pH and
flow decline. Thus, more elevated fluoride concentrations
could result, including those in the saliva, and more
fluoride would be available to both erupted and developing
teeth. Moore et al. (46), in confirmation of earlier reports, found the acid-base balance of infants to vary
directly with dietary acid load. The acid-base status of
the subjects ranged from normal to severe acidosis depending
on whether mother's milk or a formulation was fed.. If the
pH-dependence of renal fluoride metabolism in infants and
children is as important as indicated by our studies with
rats, dogs, and human adults, then fluoride balance and the
ensuing dental effects in such populations fed a standard
amount of fluoride should vary widely. Because of this
potentially large effect of diet on fluoride excretion, as
well as the difficulty of determining the absorbable
fraction of the fluoride in various infant and toddler
dietary preparations, the attempts to establish "optimal
fluoride intake" based only on age may have little practical
meaning.

References and Notes

1. Chen, P.S., Jr., Smith, F.A., Gardner, D.E., O'Brien, J.A., and Hodge, H.C. (1956). Renal clearance of fluoride. Proc. Soc. Exptl. Biol. Med. 92; 879-883.

2. Carlson, C.H., Armstrong, W.D., Singer, L., and Hinshaw, L.B. (1960). Renal clearance of radiofluoride in the dog. Am. J. Physiol. 198; 829-832.

3. Carlson, C.H., Armstrong, W.D., and Singer, L. (1960). Distribution and excretion of radiofluoride in the human. Proc. Soc. Exptl. Biol. Med. 104; 235-239.

4. Walser, M., and Rahill, W.J. (1966). Renal tubular transport of fluoride compared to chloride. Am. J. Physiol. 210; 1290-1292.

5. Whitford, G.M., Pashley, D.H., and Stringer, G.I. (1976) Fluoride renal clearance: a pH-dependent event. Am. J. Physiol. 230; 527-532.

6. Burg, M., Stoner, L., Cardinal, J., and Green, N. (1973) Furosemide effect on isolated perfused tubules. Am. J. Physiol. 225; 119.

7. Rector, F.C., Jr. (1971). Renal secretion of hydrogen. In: The Kidney: Morphology, Biochemistry and Physiology, edited by C. Rouiller and A.F. Muller, New York: Academic Press, vol. III, p. 235.

8. Harmon, C., Whitford, G.M., Pashley, D., Reynolds, K., and Stock, H. (1976). pH-dependence of fluoride renal clearance in dogs. Soc. Exptl. Biol. Med. (Southeastern Section), Abstract #13.

9. Malnic, G., Aires, M. de M., Giebisch, G. (1972). Micropuncture study of renal tubular hydrogen ion transport in the rat. Am. J. Physiol. 222; 147-158.

10. Pitts, R.F. (1974). Physiology of the Kidney and Body Fluids. Chicago: Year Book Med. Pub., pp. 217-221.

11. Whitford, G.M., Pashley, D.H., and Reynolds, K.E. (1977) Fluoride absorption from the rat urinary bladder: a pH-dependent event. Am. J. Physiol. 232; F10-F15.

12. Koss, L.G. (1969). The asymmetric unit membrane of the epithelium of the urinary bladder of the rat. Lab. Invest. 21; 154-168.

13. Maude, D.L. (1974). Mechanisms of tubular transport of salt and water. In: MTP Intern Rev. Sci. Kidney and Urinary Tract Physiology, edited by A.C. Guyton and K. Thurau. London: Butterworths, ser. 1, vol. 6, p.39-78.

14. Borzelleca, J.F. (1965). Studies on the mechanisms of drug movement from the isolated urinary bladder. J. Pharmacol. Exptl. Therap. 148; 111-116.

15. Fellows, G.J., and Marshall, D.H. (1972). The permeability of human bladder epithelium to water and sodium. Invest. Urol. 9; 339-344.

16. Hlad, C.J., Jr., Nelson, R., and Holmes, J.H. (1956). Transfer of electrolytes across the urinary bladder in the dog. Am. J. Physiol. 184: 406-411.

17. Levinsky, N.G., and Berliner, R.W. (1959). Changes in composition of the urine in ureter and bladder at low urine flow. Am.J. Physiol.196; 549-553.

18. Maluf, N.S.R. (1955). Further studies on absorption through the human bladder. J. Urol. 73; 830-835.

19. Milroy, E.J.G., Cockett, A.T.K., and Roberts, A.P. (1974) The bladder lymphatics: a study of drug transport. Invest. Urol. 12; 69-73.

20. Rosenfeld, J.B., Aboulafia, E.D., and Schwartz, W.B. (1963). Influence of nonionic diffusion on absorption of NH_4^+ and HCO_3^- from the bladder. Am. J. Physiol. 204; 568-572.

21. Whitford, G.M., and Pashley, D.H. (1974). Soft tissue distribution of fluoride. J. Dent. Res. 53 Special Issue, Abstract # 705.

22. Carlson, C.H., Singer, L., Armstrong, W.D. (1960). Radiofluoride distribution in tissues of normal and nephrectomized rats. P.S.E.B.M. 103; 418-420.

23. Barzel, U.S., and Jowsey, J. (1969). The effects of chronic acid and alkali administration on bone turnover in adult rats. Clin Sci. 36; 517-524.

24. Albright, F., and Reifenstein, F.G., Jr. (1948). In: The Parathyroid Glands and Metabolic Bone Disease. Williams and Wilkins Co., Baltimore. pp. 241-247.

25. Glassman, V.P., Safirstein, R., and DiScala, V.A. (1974). Effects of metabolic acidosis on proximal tubule ion reabsorption in the dog kidney. Am. J. Physiol. 227; 759-765.

26. Smith, F.A. (1966). Metabolism of inorganic fluoride. In: Handbook of Experimental Fluoride, Pharmacology of Fluorides, edited by F.A. Smith. Berlin: Springer-Verlag, vol. 20(1); pp. 53-140.

27. Messer, H.H., Armstrong, W.D., and Singer, L. (1973). Fluoride, parathyroid hormone, and calcitonin: interrelationships in bone calcium metabolism. Calc. Tis. Res. 13: 217-225.

28. Messer, H.H., Armstrong, W.D., and Singer, L. (1973). Fluoride, parathyroid hormone, and calcitonin: effects on metabolic processes involved in bone resorption. Calc. Tis. Res. 13; 227-233.

29. Weidmann, S.M., and Weatherell, J.A. (1970). Distribution in hard tissues. In: Fluorides and Human Health, Chapt. 4 (part 3), p. 104-128. World Health Organization Monograph Series #59. WHO, Geneva.

30. Heisler, N. (1975). Intracellular pH of isolated rat diaphragm with metabolic and respiratory changes of extracellular pH. Resp. Physiol. 23; 243-255.

31. Tung, S.H., Bettice, J., Wang, B.C., and Brown, Jr., E.B. (1976). Intracellular and extracellular acid-base changes in hemorrhagic shock. Resp. Physiol. 26; 229-237.

32. Ellis, D., and Thomas, R.C. (1976). Direct measurement of the intracellular pH of mammalian cardiac muscle. J. Physiol. 262; 755-771.

33. Levin, G.E., Collinson, P., and Baron, D.H. (1976). The intracellular pH of human leucocytes in response to acid-base changes in vitro. Clin. Sci. Mol. Med. 50; 293-299.

34. Borei, H. (1945). Inhibition of cellular oxidation by fluoride. Arkiv fur Kemi, Mineralogi o. Geologi 20A(7); 1-215.

35. Lent, B., Conroy, C.W., Erman, J. (1976). The effect of ionic strength on the kinetics of fluoride binding to cytochrome c peroxidase. Archiv. Biochem. Biophys. 177; 56-61.

36. Waddell, W.J., and Bates, R.G. (1969). Intracellular pH. Physiol. Rev. 49; 286-324.

37. Waddell, W.J., and Butler, T.C. (1959). Calculation of intracellular pH from the distribution of 5,5-dimethyl-2,4-oxazolidinedione (DMO). Application to the skeletal muscle of the dog. J. Clin. Invest. 38; 720-729.

38. Irvine, R.O.H., Saunders, S.J., Milne, M.D., and Crawford, M.A. (1960). Gradients of potassium and hydrogen ion in potassium-deficient voluntary muscle. Clin. Sci. 20; 1-18.

39. Boron, W.F., and Roos, A. (1976). Comparison of microelectrode, DMO, and methylamine methods for measuring intracellular pH. Am. J. Physiol. 231; 799-809.

40. Miyao, K. (1967). The measurement of intracellular pH by DMO method and the buffering capacity of the tissues after acid infusion. J. Physiol. Soc. Japan 29; 18-28.

41. Schloerb, P.R., and Grantham, J.J. (1965). Intracellular pH measurement with tritiated water, carbon-14 labeled 5, 5-dimethyl-2, 4-oxazolidinedione, and chloride-36. J. Lab. Clin. Med. 65; 669-676.

42. Carlson, C.H., Armstrong, W.D., and Singer, L. (1960). Distribution, migration, and binding of whole blood fluoride evaluated with radiofluoride. Am. J. Physiol. 199; 187-189.

43. Wallace-Durbin, P. (1954). The metabolism of fluorine in the rat using ^{18}F as a tracer. J. Dent. Res. 33; 789-800.

44. Lu, F.C., Mazurkiewicz, I.M., Grewal, R.S., Allmark, M.G. (1965). Acute toxicity of sodium fluoride for Rhesus monkeys and other laboratory animals. Acta Pharmacol . et Toxicol. 22: 99-106.

45. Lu, F.C., Mazurkiewicz, I.M., Grewal, R.S., Allmark, M.G. and Boivin, P. (1961). Effect of sodium fluoride on responses to various central nervous system agents in rats. Toxicol. Appl. Pharmacol. 3; 31-38.

46. Moore, A., Ansell, C., and Barrie, H. (1977). Metabolic acidosis and infant feeding. Brit. Med. J. 1; 129-131.

The authors gratefully acknowledge the assistance of Drs. J.G. Weatherred and J.M. Ginsburg. Radiofluoride was manufactured and supplied under the supervision of Mr. Robert Kirkland at the Georgia Institute of Technology and supported by E.R.D.A. Reactor Sharing Program, Contract B 380. This work was supported by National Institute of Dental Research Grant DE-04332.

9

Fluoride Resistance
in Cell Cultures

Mary G. Repaske and John W. Suttie

The biological effects of excessive fluoride have been studied at a number of levels of cellular order and complexity in an attempt to understand the metabolic alterations resulting from an increased cellular concentration of fluoride. Studies of the effects of fluoride on cultured mammalian cells were first initiated (Carlson and Suttie, 1967; 1967a) in our laboratory because of our investigations of the metabolic effects of fluoride ingestion in intact animals. The observation of (Carlson and Suttie, 1966; Zebrowski and Suttie, 1966) a number of alterations of carbohydrate and lipid metabolism in rats fed fluoride led to concern that many of these might be secondary responses to an altered pattern of dietary intake in these animals. Even when rather elaborate measures were taken to minimize the effects of altered food intake (Suttie, 1968), it was still a potential problem. By using cultured cells, we hoped to be able to look at some of the same areas of metabolism under conditions where changes in food intake were not a factor. Although there has been a report (Berry and Trillwood, 1963) that the growth of cultured mammalian cells is sensitive to as little as 0.1 ppm F (ca. 5 µM) in the growth media, other investigators have not been able to confirm this finding (Carlson and Suttie, 1967; Armstrong et al., 1965; Albright, 1964; Nias, 1965; Pace and Elrod, 1960; Hongslo et al., 1974). Our experience utilizing either HeLa cells (human epithelial carcinoma) grown in monolayers (Carlson and Suttie, 1967; 1967a) or L cells (mouse fibroblasts) grown in suspension culture (Drescher and Suttie, 1972; Quissell and Suttie, 1972; 1973) has been that the growth of these cells, measured by an increase in either cell number or amount of cellular protein, is demonstrably affected by 20 ppm F (ca. 1 mM) in the growth media, and that growth ceases almost completely at 40 ppm F (Figure 1).

The intracellular fluoride concentrations associated with these concentrations of fluoride in the medium have been

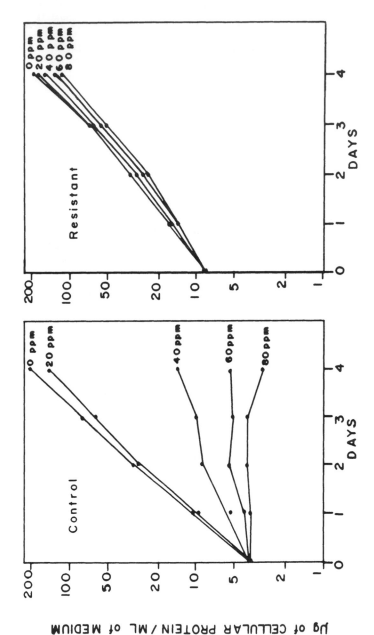

Fig. 1. Effect of fluoride concentration on the growth of the normal strain of L cells (left) and the fluoride-resistant strain of L cells (right). The cells were grown from an initial cell density of 2 x 10⁴ cells/ml in 500 ml siliconized florence flasks at 37° C. Growth was measured as an increase in cellular protein. For details see Quissell and Suttie (1972).

measured (Drescher and Suttie, 1972), and it has been shown that the ratio of intracellular to extracellular fluoride (F_{in}/F_{out}) is 0.3-0.4. This value is rather independent of the fluoride concentration in the media and of the type of cell utilized. This means that the intracellular fluoride concentrations that are associated with a significant decrease in the growth rate of these cells are in the range of 0.5 mM, and that it is difficult to demonstrate any effect on growth of these cells at intracellular fluoride concentrations less than 0.25 mM. These fluoride concentrations are considerably higher than those found in plasma and soft tissue of animals fed toxic amounts of fluoride. The plasma of rats fed very high levels of fluoride in the diet (450-600 ppm) has been observed to reach a level of from 0.1-0.15 mM (Shearer and Suttie, 1967; Simon and Suttie, 1968), and soft tissue concentrations are somewhat lower than plasma concentrations. The significance of these concentrations depends on the metabolic pathway being considered. The sensitivity of various enzymes to fluoride varies widely (Cimasoni, 1970; Wiseman, 1970), ranging from 0.5 to 1 mM F for some divalent cation-requiring enzymes such as enolase to as little as 20 μM F for some very sensitive esterases which show substantial fluoride inhibition at this level (Haugen and Suttie, 1974). This wide range of fluoride sensitivity would suggest that a number of metabolic activities might be inhibited when cultured cells are exposed to fluoride, and that cultured mammalian cells might be a useful system for determining some of the metabolic effects of toxic amounts of fluoride.

Fluoride-resistant Bacterial Strains

Several bacterial and mammalian cell lines have been adapted to grow at levels of fluoride normally inhibiting to growth. This adaptation has usually been achieved by repeatedly transferring the cells to media containing successively high levels of fluoride. For instance, Green and Dodd (1957) adapted oral lactobacilli to grow in as much as two percent sodium fluoride (ca. 10,000 ppm F). Wiggert and Werkman (1939) also grew P. pentosaceum in up to 0.02 M NaF (ca. 400 ppm F). Volk (1954) demonstrated that these resistant cells were inhibited by fluoride after lyophilization and suggested that the F resistance was due to decreased permeability of cells to sodium fluoride. The mechanism of this permeability was not investigated. S. faecalis have also been adapted to grow in high levels of fluoride (Williams, 1967). These resistant cultures grew slower and produced less acid than the parent strain. Resistance in these cells was only a phenotypic expression, as the cells lost their resistance when they were recultured in a fluoride-free medium. The mechanism of fluoride resistance appeared to be at least two-fold; the resistant cells were slightly less permeable to fluoride than was the parent strain (Williams, 1968), and data on the ratios of the concentration of 2- and 3-phosphoglyceric acid to the concentration of phosphoenolpyruvate suggested that a fluoride-insensitive form of enolase may have existed in the resistant cells (Williams, 1969).

Hamilton (1969) investigated a genotypic resistance of S. salivarius to fluoride. Cells which were adapted to growth in up to 14 mM NaF (ca. 250 ppm F) produced less acid per unit glucose utilized than the parent strain. In contrast to the phenotypically resistant S. faecalis strains, these cells retain their resistance to fluoride even after being cultured for 500 generations in the absence of fluoride. The mechanism of this fluoride resistance is unclear, but it is known that the primary site of fluoride action in the parent strain is on glucose transport via a decrease in phosphoenolpyruvate levels required for this transport system (Kanapka and Hamilton, 1971). In resistant cells grown in sodium fluoride at pH 7.2, the phosphoenolpyruvate levels were similar to those in controls. However, when the pH of the growth medium was lowered to 5.8, glucose utilization and enolase activity of the resistant cells was inhibited to the extent that transport activity was impaired. Hamilton (1977) interpreted this data to mean that membrane impermeability to fluoride was not the mechanism of resistance.

Fluoride-resistant Mammalian Cells

The observations in bacterial systems demonstrate that organisms can become resistant to the effects of fluoride, and that resistance may develop in different ways. It is, therefore, not surprising that several fluoride-resistant mammalian cell lines also have been developed. Carlson and Suttie (1967) developed a fluoride-resistant strain of HeLa cells (monolayer) which were adapted to grow in 95 ppm F (5 mM). This strain showed a much slower growth rate than did the controls, even in fluoride-free medium, but the mechanism of fluoride resistance was not investigated. Hongslo et al. (1974) adapted a LS strain of L cells to grow in up to 24 mM sodium fluoride. Cells adapted to 6 mM sodium fluoride (ca. 110 ppm F) were studied in some detail, and the growth rate of these cells was similar to that of normal cells. The resistance apparently involved a genetic alteration since adapted cells removed from fluoride were still resistant two years later. Again the means of achieving resistance was not studied.

Only one fluoride-resistant mammalian cell line has been extensively investigated. Quissell and Suttie (1972)

developed a fluoride-resistant strain of L cells (mouse
fibroblasts) by increasing the fluoride concentration in the
growth medium by 5 ppm each month until a maximum of 70 ppm F
was reached. The growth rate of these fluoride-resistant
cells in the absence of fluoride was similar to that of the
original fluoride-sensitive strain, and growth was affected
very little in the resistant cells by the presence of sufficient fluoride (40 ppm or ca. 2 mM) to almost completely
inhibit the growth of sensitive cells (Figure 1). The adaptation of these cells to growth in the presence of high
fluoride concentrations involved a genotypic alteration, and
cells retained their resistance to fluoride after being grown
for 80 generations in fluoride-free medium. It was subsequently shown that it was not necessary to select the resistant cells by gradually increasing the fluoride concentration
of the medium; they could be obtained by cloning cells directly in media containing 70 ppm F. When this was done, it was
found that mutants capable of growing in this concentration
of fluoride were present in the normal population in a ratio
of about 1 per 10^6 nonresistant cells. Although there
appeared to be some differences in the characteristics of
different fluoride-resistant strains selected by this method,
these differences were not investigated.

A comparison of various physical and metabolic properties of the resistant and normal cells did not indicate any
major differences between the two strains. Cell morphology
and size were unchanged by resistance. Glucose metabolism
did not seem to be greatly altered, as the rate of glucose
utilization in the two strains of cells was similar (Table 1).
The intracellular sodium and potassium concentrations did not
differ significantly between the two cell lines, and chloride,
bromide, and iodide were distributed similarly in resistant
and sensitive lines (Table 2).

The major difference between the fluoride-resistant and
the sensitive cells was shown to be a decreased intracellular
fluoride concentration. At a medium fluoride concentration
of 70 ppm, the ratio of intracellular to extracellular
fluoride was 0.05 for the resistant cells compared to 0.27
for the sensitive cells (Table 2). This decreased intracellular fluoride could result either from an impermeability of
the cell membrane to fluoride or from an active transport of
fluoride out of the cell into the medium. To differentiate
between these two possibilities, cells were incubated at
various temperatures, and the fluoride distribution ratio
(F_{in}/F_{out}) was determined. When cells were incubated at $0°$ C,
the intracellular fluoride level rose, but when the temperature of incubation was raised to $37°$ C, fluoride was pumped
out of the cell against a concentration gradient (Table 3).

Table 1. Utilization of glucose by control and fluoride-resistant L cells.

Fluoride conc. (ppm)	Glucose utilization µmole/mg protein		Glucose utilization rate µmole/h/mg protein	
	Control	Resistant	Control	Resistant
0	15	17	0.53	0.46
20	19	19	0.64	0.48
40	140	13	-	0.53
60	377	19	-	0.48

Glucose utilization was measured over a 4-day growth period either as total used per milligram protein accumulated (glucose utilization) or as a differential rate of gluocse utilization which takes into account differences in growth rate (glucose utilization rate). The growth rate of the control cells at 40 and 60 ppm F is too slow to allow a meaningful calculation of the latter value. The values are means of 6 determinations, and the S.E. was from 10-15% of the mean. For details see Quissell and Suttie (1972).

Table 2. Distribution of inorganic ions in control and fluoride-resistant cells.

Ion	Control cells	F-resistant cells
	Distribution ratio ($conc_{in}/conc_{out}$)	
Fluoride (^{18}F)	0.27	0.05
Chloride (^{36}Cl)	0.31	0.33
Bromide (^{82}Br)	0.30	0.33
Iodide (^{125}I)	0.40	0.35
	Intracellular concentration (mM)	
Sodium	15	30
Potassium	158	150

The fluoride-resistant cells were grown in 70 ppm F. Values are means of 4 determinations (S.E. for measurements was about 5% of the mean). For details see Quissell and Suttie (1972).

Table 3. Effect of temperature on fluoride distribution in fluoride-resistant L cells.

Experimental conditions	Ratio F_{in}/F_{out}
Incubation at 37° C	0.06 ± 0.06
Incubation 1 h at 0° C	0.19 ± 0.01
Incubation 3 h at 0° C	0.33 ± 0.01
Incubation 3 h at 0° C, then at 37° C	< 0.01 (4 exp.)

Values are mean $\pm$ S.E. for 4 determinations. Cells were incubated in 10 ppm F containing ^{18}F. For details see Quissell and Suttie (1972).

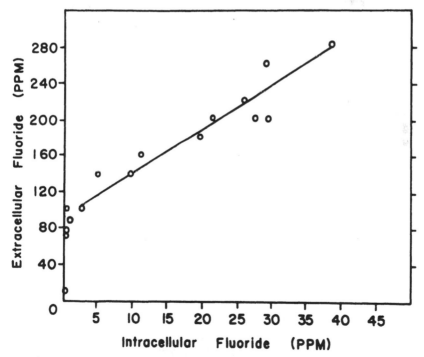

Fig. 2. Effect of fluoride concentration on cellular distribution of fluoride in fluoride-resistant L cells.

The cells were incubated in radioactive fluoride and the concentrations calculated from the distribution ratios observed. For details see Quissell and Suttie (1972).

Incubation of resistant cells in ouabain prevented the resistant cells from maintaining a low intracellular fluoride level which again suggested that the mechanism of fluoride resistance was an active transport of fluoride out of the cells. If an "active" process is involved, the transport system should be saturable. Resistant cells were therefore grown in increasingly higher levels of fluoride (from 70-200 ppm F), and both growth and fluoride distribution ratios studied. At medium fluoride concentrations above 100 ppm, growth of resistant cells was progressively inhibited, and the intracellular fluoride level rose (Figure 2). These data indicated that the fluoride pump of the resistant cells could be saturated, after which fluoride levels inside the cells would increase, and an inhibition similar to that seen in normal cell strains would occur. This study clearly demonstrated that the mechanism of fluoride resistance in this L cell strain involved an active pumping of fluoride out of the cell in order to maintain a low intracellular fluoride concentration in the presence of high extracellular fluoride levels. However, the possibility was not completely ruled out that a change in membrane permeability to fluoride (which has been suggested for many bacterial strains) could account for part of the resistance. Neither was there any evidence obtained which would indicate how the fluoride pump functioned. The rather high frequency of this mutation in the normal cell population might suggest that an alteration in a single protein somehow converts an existing transport system into systems more specific for fluoride. Investigations of this hypothesis have, however, not been undertaken.

Metabolic Effects of Fluoride

The metabolic basis for inhibition of cell growth in fluoride is unknown. However, some insight into this problem has been gained by a series of studies carried out in our laboratory (Carlson and Suttie, 1967a; Quissell and Suttie, 1973; Drescher, 1974) on the metabolic changes occurring in cells treated with fluoride. Potential sites at which fluoride could inhibit cell growth include a direct interference in protein, DNA, or RNA synthesis. Studies by Drescher (1974) on the incorporation of radioactive precursors of these macromolecules into L cells cultured at 30 ppm F revealed that DNA synthesis, protein synthesis, and RNA synthesis were inhibited to varying degrees. However, the onset of the inhibition of synthesis was simultaneous for all three macromolecules and maximal inhibition of DNA, protein, and RNA synthesis occurred within 30-60 minutes of fluoride addition. These results suggest that the inhibition of macromolecular synthesis is a secondary consequence of a primary action of fluoride elsewhere in cellular metabolism, or that multiple primary effects of fluoride on DNA, RNA, and protein synthesis occur at the same time. The first alternative seems to be the more reasonable one.

Whatever aspects of cellular metabolism are influenced by fluoride, it presumably acts through selective enzyme inhibition. Enolase, a glycolytic enzyme inhibited in vitro by fluoride, is a prime candidate in cellular effects of fluoride. Rabbit muscle enolase has a K_i for F^- inhibition in the presence of Mg^{2+} and PO_4^{-3} of 9.5 ppm (0.5 mM) F (Cimasoni, 1972). This concentration corresponds to the intracellular level of fluoride found in L cells grown in the presence of 30 ppm (ca. 1.5 mM) F. Drescher (1974) has studied enolase inhibition in these cells by determining the steady-state levels of 3-phosphoglyceric acid (3-PGA), 2-phosphoglyceric acid (2-PGA), phosphoenolpyruvate (PEP), and pyruvate in both control cells and in cells grown in from 5 to 30 ppm F (Table 4). If an enzyme (enolase) is inhibited, an accumulation of substrate (2-PGA) relative to the amount of product (PEP) formed should occur. In these experiments, the concentrations of 2-PGA and 3-PGA increased as the level of fluoride in the medium rose, but the PEP and pyruvate levels were not dramatically altered. The increased ratio (2-PGA/ PEP) provides evidence that enolase was inhibited, but the significance of this in terms of an effect of fluoride on cellular growth is unclear. Enolase is not normally a rate-limiting enzyme in glycolysis, and the data indicate that there was an effect on enolase at fluoride levels of 5 and 10 ppm but this effect was not associated with decreased cell growth.

These experiments, which measured steady-state concentrations of metabolites, were followed by studies on the flux of these metabolites through the glycolytic pathway. Uniformly labeled ^{14}C-D-glucose was added to cells grown in both the presence and absence of 30 ppm F, and the amount of ^{14}C-lactate formed at various times was determined. The total amount of ^{14}C-lactate produced in fluoride-treated cells over different time intervals was decreased, as there were fewer cells in fluoride-treated cultures. When the data were analyzed in terms of ^{14}C-lactate produced per μg protein (or per cell) per hour, it appeared that the exposure of L cells to 30 ppm F caused a 20-25% inhibition of glycolytic flux during the early stages of cell growth. However, the relationship of this inhibition of glycolytic flux (only 25%) to the 50% growth inhibition seen with 30 ppm F is unknown, as is the extent to which the energy required for growth in these cells is obtained from a source other than glycolysis.

Table 4. Effect of fluoride on the intracellular level of glycolytic intermediates in L cells.

| Treatment | Concentration of metabolite | | | | $\frac{\text{3-PGA}}{\text{2-PGA}}$ | $\frac{\text{2-PGA}}{\text{PEP}}$ |
	3-PGA	2-PGA	PEP	Pyruvate		
Control	0.26	0.07	0.12	0.06	3.9	0.6
4 ppm F	0.53	0.06	0.10	0.06	8.8	0.6
10 ppm F	1.2	0.14	0.15	0.06	8.9	0.9
20 ppm F	3.3	0.36	0.14	0.06	9.2	2.6
30 ppm F	7.8	0.89	0.21	0.04	8.7	4.2

Values are μmole of 3-PGA, 2-PGA, and PEP per gram of cell protein and μmole of pyruvate per equal amounts of cell pellet. The values are the means of 3 or 4 cultures (S.E. for most measurements was less than 10% of the mean). For details see Drescher (1974).

Analysis of the adenosine triphosphate concentrations in fluoride-treated cells indicates that the inhibition of glycolitic flux does not translate into lower cellular ATP levels. Intracellular ATP levels were determined at various times after the addition of 30 ppm F to L cell cultures (Drescher, 1974). During the logarithmic growth phase, both control and fluoride-treated cells experienced a similar fall in intracellular ATP levels, from 20 μmoles/g cell protein to 10 μmoles/g cell protein. Also, both fluoride-treated and control cells maintained similar intracellular sodium and potassium concentrations. It appears, therefore, that there is no obvious correlation between cellular ATP levels, inhibition of enolase, glycolytic flux, and growth. It should be considered, however, that the cells may adjust their growth to maintain a "normal" ATP concentration and that this would mask any effect of fluoride on ATP production.

One aspect of L cell metabolism that is significantly altered by fluoride is diphosphopyridine nucleotide metabolism. The fluoride effect is twofold: (1) Within two hours of fluoride addition (30 ppm), the NADH/NAD ratio rises from 0.21 in control cells to a new constant value of about 0.30; (2) In addition, within 12 hours, the total cellular concentration of NAD + NADH decreased to a new constant level (50% that of controls) in fluoride-treated cells (Table 5). How fluoride acts to either directly or indirectly achieve this 50% reduction in pyridine nucleotide concentration is undetermined. However, the reduction is not merely a cellular response to a slower growth rate. We have recently found (Repaske and Suttie, unpubl. data) that ouabain, sodium oxamate, and incubation at a decreased temperature all decrease cellular growth with no significant effect on pyridine nucleotide levels.

Factors Influencing the Response of Cells to Fluoride

A second approach to the determination of the site of fluoride inhibition in cultured cells is to analyze the effects of various metabolites which relieve the inhibition. Several metabolites have been shown to improve the growth of fluoride-treated cells. The addition of sodium pyruvate (0.3 mM) to Murine Leukemic Lymphoblasts was shown by Albright to result in a 60% improvement in the growth of cells grown in 20 ppm F (Albright, 1966). Similarly, the growth inhibition produced by 30 ppm F in L cells can be reversed by almost 50% when 10 mM sodium pyruvate is added to the medium (Drescher, 1974). It has been suggested that the addition of pyruvate to fluoride-treated cells permits reoxidation of NADH via the conversion of pyruvate to lactate, thus bypassing the (inhibited) enolase reaction. It has,

Table 5. Effect of fluoride on intracellular levels of NAD and NADH in L cells.

Measurement	12 h Incubation		24 h Incubation	
	Control	30 ppm F	Control	30 ppm F
NAD (μmole/g protein)	3.20	1.40	2.90	1.50
NADH (μmole/g protein)	0.54	0.41	0.60	0.45
Total NAD + NADH	3.74	1.81	3.50	1.95
NADH/NAD	0.17	0.29	0.21	0.30

Fluoride was added to logarithmically growing cultures and the amount of NAD and NADH determined at 2, 12, 24, and 48 hours. The concentrations did not change appreciably at the different times, and the data at 2 and 48 hours are similar to those shown. The values are the average of two different cultures which differed by less than 10%. For details see Drescher (1974).

Table 6. Effect of fluoride, pyruvate, and glutamine on L cell pyridine nucleotides.

Treatment	μmole Pyridine nucleotide/10^{10} cells			$\dfrac{\text{NADH}}{\text{NAD}}$
	NAD^+	NADH	Total	
Control	8.4	1.6	10.0	0.19
35 ppm F	5.4	1.4	6.8	0.27
F + 6 mM glutamine	5.3	1.4	6.6	0.26
F + 5 mM pyruvate	6.3	1.3	7.6	0.21

Pyridine nucleotide content of the cells was determined 16 hours after all additions. In separate experiments the addition of glutamine on pyruvate alone had no effect on pyridine nucleotide concentration. Values are an average of duplicate cultures which differed by less than 10%.

however, not been possible to see the expected decrease in pyruvate in fluoride-treated cells. What can be shown is that increasing the concentration of pyruvate in the medium partially readjusts the pyridine nucleotide levels in fluoride-treated cells. The addition of 5 mM sodium pyruvate to L cells grown in 35 ppm F improved growth about 25% and also elevated pyridine nucleotide levels about 25% (Table 6). How pyruvate is affecting this increase in pyridine nucleotide concentration is undetermined.

Several amino acids have also been shown to improve the growth of fluoride-treated cells (Figure 3). Serine and glycine, both of which can be metabolized to pyruvate, significantly decrease the fluoride inhibition. Asparagine and glutamine also decrease the fluoride effect, while glutamic acid and aspartic acid do not (Drescher, 1974). These data must be analyzed in terms of the effect of these amino acids on control cell growth. Drescher followed up these initial observations with a detailed investigation of the effect of glutamine. Ammonium ion, at low nontoxic levels, caused a slight improvement in growth, suggesting that the glutamine effect may be partially accounted for by NH_3 produced during glutamine metabolism. Conversely, carboxy-pyrrolidone, formed by the spontaneous decomposition of glutamine in the medium, and α-ketoglutarate, the metabolite of glutamine which can be utilized for energy metabolism, were not capable of improving growth of fluoride-treated cells. Further studies (Drescher, 1974) comparing the metabolism of ^{14}C-glutamine in both control and fluoride-treated cells also suggested that the carbon skeleton of glutamine is not involved. These data suggest that the γ-amide of glutamine is the important factor in relieving the growth inhibition produced by fluoride. What the γ-amide does is still unclear. Glutamine functions as a γ-amide donor in many synthetic reactions. One example is the donation of the amide group in diphosphopyridine nucleotide synthesis. Drescher suggests that glutamine might be overcoming the fluoride inhibition via an increase in the level of NAD + NADH. However, subsequent studies (Table 6) indicated that glutamine had no effect on either the 50% decline in total cellular pyridine nucleotide concentration or on the NADH/NAD ratio. Glutamine does have a slight effect (Drescher, 1974) on the cellular distribution of fluoride. Increasing the glutamine concentration at a given fluoride concentration results in a slight decrease in intracellular fluoride. This effect does not, however, appear to be sufficient to account for the overall benefit seen with high concentrations of glutamine in the medium.

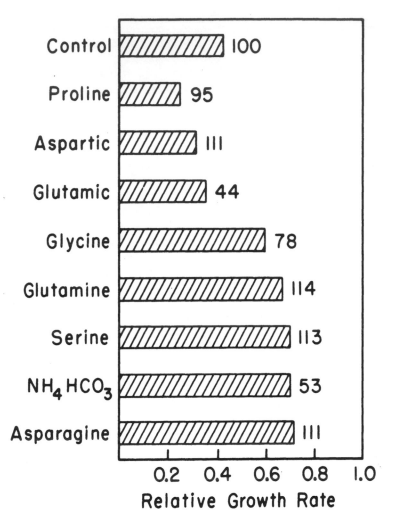

Fig. 3 Effect of the addition of various amino acids on the degree to which fluoride inhibits L cell growth over a 4-day period in the presence of 30 ppm F.

The various compounds were added to normal and fluoride-containing media at a concentration of 5 mM. Growth rate of the fluoride-treated cells was measured relative to their control. The number above each bar is the growth rate (percent of growth in non-fluoride media) of the cells after addition of the compound. For details see Drescher (1974).

Summary

Cultured mammalian cells offer some advantages in the study of fluoride effects on metabolism. The effects seen are likely to apply to toxicity in intact animals, and many of the problems associated with metabolic studies in intact animals are alleviated. These cells have been particularly useful in assessing the nature of fluoride resistance. A number of bacterial strains and cultured cell lines have been reported to be resistant to fluoride toxicity. In the one fibroblast cell line extensively studied, the resistance has been shown to be the result of a genetic alteration which has given the cell mutant the ability to maintain low intracellular fluoride concentrations in the face of high concentrations of fluoride in the medium.

The basic metabolic lesions which are responsible for the toxic effects of fluoride on cultured cells have not yet been identified. Studies of metabolic alterations within these cells, and studies of factors which influence fluoride toxicity have, however, led to the beginning of an understanding of the metabolic consequences of excessive fluoride.

Acknowledgements

Research supported by the College of Agricultural and Life Sciences, University of Wisconsin-Madison, in part by Grant No. Am-15521 from the National Institutes of Health, and in part by U.W. Experiment Station Project 809.

References

Albright, J. A. (1964) Inhibitory levels of fluoride on mammalian cells. Nature 203; 976.

Albright, J. A. (1966) Effect of fluoride on mammalian cells: partial reversal by pyruvate. J. Oral Therap. and Pharm. 2; 436-439.

Armstrong, W. D., C. H. Blomquist, L. Singer, M. E. Pollock, and L. C. McLaren (1965) Sodium fluoride and cell growth. Brit. Med. J. 1; 486-488.

Berry, R. J., and W. Trillwood (1963) NaF and cell growth. Brit. Med. J. 2; 1064.

Carlson, J. R., and J. W. Suttie (1966) Pentose phosphate pathway enzymes and glucose oxidation in fluoride-fed rats. Am. J. Physiol. 210; 79-83.

Carlson, J. R., and J. W. Suttie (1967) Effect of sodium
fluoride on HeLa cells. I. Growth sensitivity and adap-
tion. Exp. Cell Res. 45; 415-422.

Carlson, J. R., and J. W. Suttie (1967a) Effect of sodium
fluoride on HeLa cells. II. Metabolic alterations associated with growth inhibition. Exp. Cell Res. 45; 423-432.

Cimasoni, G. (1970) Fluoride and Enzymes. In Fluoride
in Medicine (T. L. Vischer, ed.), Hans Huber Pub., Bern,
pp. 14-26.

Cimasoni, G. (1972) The inhibition of enolase in vitro.
Caries Res. 6; 93-102.

Drescher, M. J. (1974) Effects of sodium fluoride on growth
and metabolism of L cells. Ph.D. Thesis, University of
Wisconsin-Madison.

Drescher, M., and J. W. Suttie (1972) Intracellular fluoride
in cultured mammalian cells. Proc. Soc. Exp. Biol. Med.
139; 228-230.

Green, G. E., and M. C. Dodd (1957) Resistance of oral lactobacilli to sodium fluoride. J. Am. Dental Assn. 54; 654-
656.

Hamilton, I. R. (1969) Growth characteristics of adapted and
ultraviolet-induced mutants of Streptococcus salivarius
resistant to sodium fluoride. Can. J. Microbiol. 15; 287-
295.

Hamilton, I. R. (1977) Effects of fluoride on enzymatic
regulation of bacterial carbohydrate metabolism. Caries
Res. 11, Suppl. 1; 262-291.

Haugen, D. A., and J. W. Suttie (1974) Fluoride inhibition
of rat liver microsomal esterases. J. Biol. Chem. 249;
2723-2731.

Hongslo, J. K., R. I. Holland, and J. Jonsen (1974) Effect of
sodium fluoride on LS cells. J. Dent. Res. 53; 410-413.

Kanapka, J. A., and I. R. Hamilton (1971) Fluoride inhibition of enolase activity in vivo and its relationship to
the inhibition of glucose-6-P formation in Streptococcus
salivarius. Arch. Biochem. Biophys. 146; 167-174.

Nias, A. H. W. (1965) Sodium fluoride and cell growth. Brit.
Med. J. 1; 1672.

Pace, D. M., and L. M. Elrod (1960) Effects of respiratory inhibitors on glucose and protein utilization and growth in strain L cells. Proc. Soc. Exp. Biol. Med. 104; 469-472.

Quissell, D. O., and J. W. Suttie (1972) Development of a fluoride-resistant strain of L cells: membrane and metabolic characteristics. Am. J. Physiol. 223; 596-603.

Quissell, D. O., and J. W. Suttie (1973) Effect of fluoride and other metabolic inhibitors on intracellular sodium and potassium concentrations in L cells. J. Cell. Physiol. 82; 59-64.

Shearer, T. R., and J. W. Suttie (1967) The effect of fluoride administration on plasma fluoride and food intake in the rat. Am. J. Physiol. 212; 1165-1168.

Shearer, T. R., and J. W. Suttie (1970) Effect of fluoride on glycolytic and citric acid cycle metabolites in rat liver. J. Nutr. 100; 749-756.

Simon, G., and J. W. Suttie (1968) Effect of method of fluoride administration on plasma fluoride concentrations. J. Nutr. 94; 511-515.

Suttie, J. W. (1968) Effect of dietary fluoride on the pattern of food intake in the rat and the development of a programmed pellet dispenser. J. Nutr. 96; 529-535.

Volk, W. A. (1954) The effect of fluoride on the permeability and phosphatase activity of Propionibacterium pentosaceum. J. Biol. Chem. 208; 777-784.

Wiggert, W. P., and C. H. Werkman (1939) Fluoride sensitivity of Propionibacterium pentosaceum as a function of growth conditions. Biochem. J. 33; 1061-1069.

Williams, R. A. D. (1967) The growth of Lancefield Group D streptococci in the presence of sodium fluoride. Arch. Oral Biol. 12; 109-117.

Williams, R. A. D. (1968) Permeability of fluoride-trained streptococci to fluoride. Arch. Oral Biol. 13; 1031-1033.

Williams, R. A. D. (1969) Glycolytic intermediates in "Fluoride-trained" and control cultures of an oral enterococcus. Arch. Oral Biol. 14; 265-270.

Wiseman, A. (1970) Effect of inorganic fluoride on enzymes. In Handbook of Experimental Pharmacology XX/2 (F. A. Smith, ed.), Springer-Verlag, New York, pp. 48-97.

Zebrowski, E. J., and J. W. Suttie (1966) Glucose oxidation and glycogen metabolism in fluoride-fed rats. J. Nutr. 88; 267-271.

10

Fluoride from Anesthetics and Its Consequences

Russell A. Van Dyke

This symposium has dealt mainly with the purposeful administration of inorganic fluoride for therapeutic or preventive purposes. It is appropriate at this time to turn the discussion to the inadvertent administration of inorganic fluoride, a situation which develops when a fluorinated drug is administered which when metabolized releases inorganic fluoride. The number of fluorinated drugs is not large, but the number of humans exposed to fluorinated drugs is because of the extensive use of fluorinated anesthetics. The following is a list of the names and chemical structures of volatile anesthetics.

Halothane	$CF_3-CClBrH$
Methoxyflurane	$CH_3-O-CF_2-CCl_2H$
Enflurane	$CF_2H-O-CF_2-CClFH$
Isoflurane	$CF_2H-O-CClH-CF_3$

All of these volatile anesthetics are metabolized by either oxidation or reduction or both. The oxidation and reduction reactions are carried out by the mixed function oxidase system, located in the endoplasmic reticulum and in other membranes of similar structure, such as the plasma or nuclear membranes. Thus, these reactions occur to the greatest extent in organs whose cells contain large amounts of these membranes, namely liver, kidneys and lungs. However, this type of enzymatic activity has been found in all tissues. The key component of this enzyme system is the hemoprotein, cytochrome P-450. The system functions as follows:

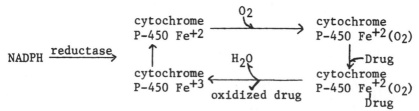

A more detailed description of this system and its function has been published elsewhere (Estabrook and Cohen, 1969, Guengench et al.,1975). The sequence of events in this system have not been precisely determined, but the important point is that in the case of an oxidation three ingredients come together on the cytochrome P-450: a pair of electrons coming from NADPH, a molecule of oxygen and the drug substrate. The result is hydroxylation of the drug substrate and formation of a water molecule.

In addition to the oxidation of drugs, cytochrome P-450 system can also carry out reduction reactions, but only if the drug substrate is capable of accepting electrons directly from cytochrome P-450. There are only a few drugs capable of this, and furthermore, these drugs must compete with oxygen, which in most cases inhibits the reduction of drugs. Thus, this reaction will only occur under conditions of lowered oxygen tension. In the case of the volatile anesthetics it has been shown definitely that at least halothane is capable of accepting an electron from this system in the absence of oxygen. It is not known at the present time whether the other volatile anesthetics will also act as electron acceptors in this reaction.

As may be expected from the structure of the volatile anesthetics, these agents undergo either an ether cleavage (also known as an O-dealkylation) or a dehalogenation. In some cases both reactions may occur. The ether cleavage occurs only oxidatively and results in the formation of an aldehyde and an alcohol. The main factor governing the rate of cleavage of the ether is the length of the alkyl group being removed. An O-methyl group is much more readily removed than is an O-ethyl group.

Since all the volatile anesthetics currently in use contain halogens, enzymatic dehalogenation may also take place (Van Dyke and Wineman, 1971; Van Dyke and Gandolfi, 1975). Contrary to some currently held beliefs, there is no direct attack on the carbon-halogen bond. Rather, oxidation of the carbon atom that carries the halogen results in an unstable molecule and leads to loss of all halogen atoms associated with the oxidized carbon atom. The rate of dehalogenation is highest when there are two halogen atoms on a terminal carbon atom. While the reaction takes place at a slower rate when the end group contains only one halogen atom, the reaction does not occur at all when the end group contains three halogen atoms (Loew et al., 1973). The metabolism of each of the volatile anesthetic agents currently in use is described below primarily from the point of view of the fluoride released rather than the significance of the other metabolites.

Methoxyflurane

The microsomal mixed-function oxidase system may select one of two sites on the methoxyflurane molecule for attack (Van Dyke and Wood, 1973), a fact that makes methoxyflurane unique amongst the volatile anesthetic agents. The methoxyflurane molecule can be attacked either at the dichloro carbon or at the ether linkage as follows:

$$CH_3-O-CF_2-CCl_2H \xrightarrow{\text{B}} CH_3-O-CF_2-COOH + 2\ Cl^-$$

$$A\downarrow \qquad\qquad\qquad\qquad\qquad \downarrow H^+$$

$$HCHO + 2F^- + HCCl_2-COOH \qquad HCHO + \underset{\underset{COOH}{|}}{COOH} + 2F^-$$

Paths A and B in the above scheme are both mediated by the cytochrome P-450 system. These reactions may be followed in vitro by measuring the appearance of inorganic fluoride rather than dichloroacetic acid or formaldehyde. While formaldehyde can be quantitated, it is slowly oxidized to carbon dioxide, thus making it a difficult product to follow, and dichloroacetic acid is not usually detected, nor quantified, at the concentrations achieved with in vitro incubations. However, through the use of the fluoride-ion-specific electrode, inorganic fluoride can be quantitated, and fluoride, once formed, does not react with other compounds in the incubation mixture. It should be kept in mind that release of inorganic fluoride is the result not of a direct attack on the carbon-fluorine bond, but rather, of an oxidation of the ether linkage causing its cleavage and secondary release of inorganic fluoride. Since an oxygen and halogen atom attached to the same carbon atoms are unstable, the halogen-carbon bond is weakened and the halogen is released as a halide. Path B is a dechlorination reaction resulting in the formation of methoxydifluoroacetic acid and inorganic chloride. Since the incubations are usually carried out in a chloride containing medium, it is difficult to quantify the inorganic chloride released from methoxyflurane. However, path B is easily followed and quantified since methoxydifluoroacetic acid breaks down to fluoride and oxalate in acid (Comereski, 1973). Therefore, the incubation is carried out at pH 7.4 and the inorganic fluoride is measured before and after acidification. The difference between the two levels of inorganic fluoride represents the amount of methoxydifluoroacetic acid produced by reaction B. This separation of the two reaction paths allows comparison as to which site on the molecule the enzyme prefers. It turns out that the amount of inorganic fluoride released from the acid-labile material is greater than that released by enzymatic cleavage of the ether linkage. Thus, reaction B, the dechlorination of methoxyflurane, is more readily accomplished than is reaction A, the ether cleavage. It should also be pointed out that at least half of the inorganic fluoride generated by the enzymatic metabolism of methoxyflurane (reaction A) enters the bones while the remainder is distributed throughout the body. In contrast, the methoxydifluoroacetic acid is probably largely removed from the blood by the kidneys where it may encounter a pH low enough (pH 5) to break it down, thereby releasing the inorganic fluoride directly in the urine and, possibly, in the kidney cells.

The possibility of a specific anesthetic-induced nephrotoxicity was first suggested by Crandell et al. (1966). They reported that renal failure with high urine volumes occurred in 13 of 41 patients who had received methoxyflurane for abdominal surgery. Urine volumes were 2.5-4.0 liters per day with a negative fluid balance, elevation of serum sodium, serum osmolality and blood urea nitrogen, and a relatively fixed urine osmolality close to that of serum. Since vasopressin administration did not reverse the high output, it was concluded that the difficulty was of renal origin and not due to ADH deficiency. The impairment lasted from 10 to 20 days in most patients. Other reports had been published previously concerning high-output renal failure in patients having received methoxyflurane anesthesia, although in these cases the toxicity had not been traced to the fact that the patients had received an anesthetic (Paddock et al., 1964; North and Stephen, 1965). It had been suggested by Taves et al. (1970) that one of the metabolites of methoxyflurane, inorganic fluoride, was responsible for the observed nephrotoxicity. They reported increased concentrations of inorganic fluoride in serum and urine of a patient who had renal dysfunction following methoxyflurane anesthesia. Frascino et al. (1970) also noted oxalic acid crystals in renal biopsy specimens as well as increased urinary oxalic acid excretion in several patients with postoperative renal insufficiency after methoxyflurane anesthesia.

Both inorganic fluoride and oxalic acid have nephrotoxic potential with the former most likely being the primary nephrotoxin. The most compelling argument for this concept was the clinical correlation of renal dysfunction and increased serum inorganic fluoride concentrations coupled with the occurrence of polyuric renal insufficiency in rats injected with sodium fluoride (Whitford and Taves, 1971). From this observation it appeared that the renal toxicity of methoxyflurane would be directly related to dose and its metabolism to inorganic fluoride. The most serious barrier was the lack of a direct cause-effect relationship of meth-oxyflurane administration and renal dysfunction. The failure to do so was a result of not being able to produce the nephrotoxicity in animals. However, Mazze et al. (1972;1973) after screening five different strains of rats found that one, the Fischer 344 strain, did develop nephrotoxicity following methoxyflurane administration. As in man, the syndrome was characterized by dose-related vasopressin resistant polyuria, hypernatremia, serum hyperosmolality and an increased serum urea nitrogen. The injection of inorganic fluoride into these animals produced changes in renal function and histologic findings similar to those seen following methoxyflurane administration.

An increase in the fluoride content of bones was used by Fiserova-Bergerova (1973) as a measure of the in vivo biodegradation of methoxyflurane. Skeletal fluoride concentrations were determined in mice and rats following interperitoneal injection of methoxyflurane on five consecutive days. The results indicated that 5.6% of the original fluoride administered as methoxyflurane was deposited in bones. It has been estimated that the amount of fluoride deposited in bone approximates the amount excreted in the urine.

The mechanism for renal toxicity of inorganic fluoride is not known. However, several theories have been proposed, such as; interference with sodium transport at the proximal convoluted tubule, inhibition of the generation of mito-chondrial ATP, and interference with the action of anti-diuretic hormones through the inhibition of adenocyclase, which affects the counter-current system of the kidney by acting as a potent vasodilator.

Humans are capable of metabolizing large amounts of methoxyflurane (Holaday et al., 1970) and because of the high lipid solubility of this agent, considerable amounts of the anesthetic remain in the body for long periods of time after anesthesia, and as might be expected the greater amount of fat the body contains, the greater amount of agent is taken up. Thus, the facts that it accumulates to a high extent in lipids and that it is easily metabolized mean a considerable level of inorganic fluoride can result from exposure to this anesthetic. What are the levels reached following anesthesia and is there a threshold level for inorganic fluoride nephrotoxicity? Cousins and Mazze (1970) tested the urine concentrating ability of the kidneys in surgical patients anesthetized with methoxyflurane by comparing urinary osmolality measured after overnight dehydration of these patients before and after operation. Defects in concentrating ability were recorded only when serum inorganic fluoride levels exceeded 50 μM (1 ppm) agreeing well with the findings in rats (Whitford and Taves, 1971). Other investigators have measured inorganic fluoride levels in surgical patients following anesthesia with methoxyflurane, but have not measured urine concentrating ability, and have found that values as high as 100 μM of serum inorganic fluoride are safe (Dobkin and Levy, 1973). In recent years 50 μM has been accepted as the threshold for production of nephrotoxicity by inorganic fluoride. These levels of inorganic fluoride released from methoxyflurane metabolism are easily reached in humans with an anesthesia of 2.5 hours, or more; and in patients with high body fat content, these levels can be achieved with even shorter exposure times.

Enflurane

An examination of the structure of enflurane reveals that oxidation of any of its carbon atoms will result in release of inorganic fluoride. On this basis, therefore, one would predict that rather high amounts of inorganic fluoride might be released from this anesthetic. The studies to date indicate that this is not so. The ether linkage is stabilized by the presence of fluorine on both sides of the ether. Thus, the only point of attack would be on the terminal dihalomethyl group. The reaction is believed to occur as follows:

$$HCF_2-O-CF_2-CClFH \longrightarrow [HCF_2-O-CF_2-COOH] + Cl^- + F^-$$

In vivo studies in rats have shown that the release of inorganic fluoride from enflurane is very limited (Greenstein et al., 1975). The ratio of inorganic fluoride released by the metabolism of methoxyflurane to that released by enflurane is approximately 8:1. One reason that the metabolism of enflurane is so low relative to that of methoxyflurane is its low solubility. In microsomes it has only approximately one fifth the solubility of methoxyflurane. Investigations in vivo have revealed that the metabolism of enflurane is much more extensive than shown by the in vitro studies, an indication that solubility may control the extent to which inorganic fluoride is released from enflurane (Chase et al., 1971; Barr et al., 1974). The metabolism of enflurane to inorganic as well as organic fluoride was studied by Cousins et al. (1976) in surgical patients who received a 1.4% inhaled concentration of enflurane. Serum inorganic fluoride reached peak levels of 22.2 μM four hours after anesthesia, but dropped to approximately control levels by 48 hours. Urinary excretion of inorganic fluoride also rose and stayed

elevated for several days. A more recent report examining
the release of inorganic fluoride from enflurane in human
volunteers exposed to prolonged anesthesia with enflurane
reveals that fairly high levels of inorganic fluoride can be
attained in serum (Mazze et al., 1977). These levels were
sufficiently high to interfere with the urine concentrating
ability of the kidneys. In this study volunteers were administered 9.6 minimum alveolar concentration (MAC) hours of
enflurane. The urine concentrating ability was measured before and after exposure to the anesthetic. It was found that
the serum inorganic fluoride level peaked six hours after
enflurane anesthesia at a concentration of approximately
33 µM. Twelve hours later, at which time serum inorganic
fluoride level had decreased 21 µM, a vasopressant was administered. By the end of the vasopressant test, serum inorganic fluoride levels had decreased to approximately 8 µM.
Thus, during the 24-hour test period, the average serum inorganic fluoride level was only 15 µM. However, these subjects
showed a 25% reduction in maximum urine concentrating ability
as compared with preanesthetic values. This conflicts with
the data obtained with methoxyflurane for which defects in
concentrating ability were recorded only when serum inorganic
fluoride levels exceeded 50 µM. This raises the possibility
that the organ toxicity is related not only to the peak level
of the toxic substance, but to the length of time the organ
is exposed to high levels. In the study by Mazze et al.
(1977) the kidneys were exposed to a mean peak serum inorganic fluoride level of 33.6 µM with values above 20 µM for 18
hours. What might happen in the case of a more rapid rise to
peak and a more rapid decrease in terms of kidney toxicity is
not known. What has been shown is that these two drugs,
methoxyflurane and enflurane, can be metabolized to inorganic
fluoride to such an extent that nephrotoxicity may develop.

Halothane

The metabolism of halothane has always been considered
to be oxidative and to result in the formation of trifluoroacetic acid (Stier, 1964). The trifluoromethyl portion of
the molecule is resistant to oxidative attack owing to the
stabilization of the carbon atom by the presence of three
halogen atoms. Recent studies,however, have indicated that
halothane can undergo reductive metabolism. There is apparently a dehydro-defluorination resulting in the production of
a number of reactive metabolites. Of importance to this discussion is the fact that inorganic fluoride can be released
from halothane provided the proper conditions are present
(Van Dyke and Gandolfi, 1976). The reactions are as follows:

$$CF_3\text{-CClBrH} \xrightarrow{\;O_2\;} [\,CF_3\text{-CHO}\,] + Cl^- + Br^-$$

$$[\,CF_3\text{-CClBr}\,]$$

$$CF_2 = CClBr + F^- \qquad\qquad CF_3COOH$$

In rats, Widger et al. (1976) examined the metabolism of halothane in both high- and low-oxygen atmospheres. Animals receiving 7% oxygen showed a sevenfold increase in plasma fluoride level compared to the nonhypoxic animals.

In humans there have been reports indicating that inorganic fluoride levels are elevated following halothane anesthesia. Creasser and Stoelting (1973) suggested that serum fluoride levels were increased during the immediate post-anesthesia hours, although the differences were statistically not highly significant. On the other hand, Young et al. (1975) studying obese patients found significant increases in serum inorganic fluoride following halothane anesthesia when these levels were compared to those individuals in a nonhalothane control group. Peak fluoride levels in the obese patients under halothane anesthesia were 10.4 μM as compared to 2 μM in the nonhalothane group. These peak concentrations were maintained during anesthetic periods, but dropped off rapidly after the anesthetic had been discontinued.

It is obvious from these studies that the inorganic fluoride released from halothane is not of importance. The increases, when noted, are transient in nature, that is increasing for just a few hours, and the peak levels achieved are not high enough to produce nephrotoxicity. However, the fact that inorganic fluoride may be released from halothane may be of importance if significant amounts of inorganic fluoride from other sources are also present.

Isoflurane

Isoflurane is metabolized only to a very limited extent, much less than all the other anesthetics discussed in this paper. Dobkin et al. (1973) measured serum inorganic fluoride levels in 189 patients having received isoflurane anesthesia for three hours. They found no appreciable elevations in fluoride levels, although isolated patients did show slight increases. Mazze et al. (1974) also examined the metabolism of isoflurane in surgical patients exposed to 1.2% isoflurane for approximately four hours. Mean peak serum inorganic fluoride levels of 4.4 μM were attained six hours after anesthesia, returning to normal values within 24 hours.

Concluding Remarks

It is evident that the volatile anesthetics release inorganic fluoride to varying degrees. It is interesting to note that the most commonly used agents, halothane and enflurane, release relatively small amounts of inorganic fluoride, but nevertheless, measureable amounts have been detected and should be taken into consideration when other fluorinated drugs are administered or when kidney function is impaired. Either instance may result in considerably elevated levels of inorganic fluoride in the serum.

References

Barr, G. A., Cousins, M. J., Mazze, R. I., et al. (1974) A comparison of the renal effects and metabolism of enflurane and methoxyflurane in Fischer 344 rats. J. Pharmacol. Exp. Therap. 188;257.

Chase, R. E., Holaday, D. A., Fiserova-Bergerova, V., et al. (1971) The biotransformation of ethrane in man. Anesthesiology 35;262.

Comereski, C. R. (1973) A source of error in urinary oxalate values after methoxuflurane anesthesia. Thesis. The University of Rochester.

Cousins, M. J., Greenstein, L. R., Hitt, B. A. et al. (1976) Metabolism and renal effects of enflurane in man. Anesthesiology 44;44-53.

Cousins, M. J., and Mazze, R. I. (1970) Methoxyflurane nephrotoxicity: a study of dose-response in man. JAMA 214;91-95.

Crandell, W. B., Pappas, S. G., and MacDonald, A. (1966) Nephrotoxicity associated with methoxyflurane anesthesia. Anesthesiology 27;250.

Creasser, C., and Stoelting, R. K. (1973) Serum inorganic fluoride concentrations during and after halothane, fluroxene and methoxyflurane anesthesia in man. Anesthesiology 39; 537.

Dobkin, A. B., Kim, D., Choi, J. K., et al (1973) Blood serum fluoride levels with enflurane (Ethrane) and isoflurane (Forane) anesthesia during and following major abdominal surgery. Can. Anaesth. Soc. J. 20;494.

Dobkin, A. B., and Levy, A. A. (1973) Blood serum fluoride levels with methoxyflurane anaesthesia. Can. Anaesth. Soc. J. 20;81-93.

Estabrook, R. W., and Cohen, B. (1969) Organization of the microsomal electron transport system. In: Microsomes and Drug Oxidations. Gilette, J. R. et al., editors. Academic Press Inc., New York.

Fiserova-Bergerova, V. (1973) Changes of fluoride content in bone: an index of drug defluorination in vivo. Anesthesiology 38;345.

Frascino, J. A., Vanamee, P., and Rosen, P. P. (1970) Renal oxalosis and azotemia after methoxyflurane anesthesia. New Eng. J. Med. 283;676-679.

Greenstein, L. R., Hitt, B. A., and Mazze, R. I. (1975) Metabolism in vitro of enflurane, isoflurane, and methoxy-flurane. Anesthesiology 42;420.

Guengench, F. P., Ballou, D. P., and Coon, M. J. (1975) Purified liver microsomal cytochrome P-450: Electron accepting properties and oxidation-reduction potential. J. Biol. Chem. 250;7405.

Holaday, D. A., Rudofsky, S., and Trenhaft, P. S. (1970) The metabolic degradation of methoxyflurane in man. Anesthesiology 33;579.

Loew, G., Trudell, J., and Motulsky, H. (1973) Quantum chemical studies of the metabolism of a series of chlorinated ethane anesthetics. Mol. Pharmacol. 9;152.

Mazze, R. I., Calverly, R. K., and Smith, N. T. (1977) Inorganic fluoride nephrotoxicity: prolonged enflurane and halothane anesthesia in volunteers. Anesthesiology 46;265-271.

Mazze, R. I., Cousins, M. J., and Barr, G. A. (1974) Renal effects and metabolism of isoflurane in man. Anesthesiology 40;536.

Mazze, R. I., Cousins, M. J., and Kosek, J. C. (1972) Dose-related methoxyflurane nephrotoxicity in rats: a biochemical and pathological correlation. Anesthesiology 36;571-587.

Mazze, R. I., Cousins, M. J. and Kosek, J. C. (1973) Strain differences in metabolism and susceptibility to the nephro-toxic effects of methoxyflurane in rats. J. Pharmacol. Exp. Therap. 184;481-488.

North, W. C., and Stephen, C. R. (1965) Hepatic and renal effects of methoxyflurane in surgical patients. Anesthesiology 26;257.

Paddock, R. B., Parker, J. W., and Guadagni, N. P. (1964) The effects of methoxyflurane on renal function. Anesthesiology 25;707.

Stier, A. (1964) Trifluoroacetic acid as metabolite of halothane. Biochem. Pharmacol. 13;1544.

Taves, D. R, Fry, B. W., Freeman, R. B., et al. (1970) Toxicity following methoxyflurane anesthesia.II. Fluoride concentrations in nephrotoxicity. JAMA 214;91-95.

Van Dyke, R. A., and Gandolfi, A. J. (1975) Characteristics of a microsomal dechlorination system. Mol. Pharmacol. 11; 809.

Van Dyke, R. A., and Gandolfi, A. J. (1976) Anaerobic release of fluoride from halothane. Drug Metab. Dispos. 4;40.

Van Dyke, R. A., and Wineman, C. G. (1971) Enzymatic dechlorination of chloroethane and propanes in vitro. Biochem. Pharmacol. 20;463.

Van Dyke, R. A., and Wood, C. L. (1973) Metabolism of methoxyflurane: Release of inorganic fluoride in human and rat hepatic microsomes. Anesthesiology 39;613.

Whitford, G. A., and Taves, D. R. (1971) Fluoride-induced diuresis: Plasma concentrations in the rat. Proc. Soc. Exp. Biol. Med. 137;458-460.

Widger, L. A., Gandolfi, A. J., and Van Dyke, R. A. (1976) Hypoxia and halothane metabolism in vitro: release of inorganic fluoride and halothane metabolite binding to cellular constituents. Anesthesiology 44;197.

Young, S. R., Stoelting, R. K., Peterson, C., et al. (1975) Anesthetic biotransformation and renal function in obese patients during and after methoxyflurane or halothane anesthesia. Anesthesiology 42;451-457.

11

The Safety of Fluoride Tablets or Drops

Harold C. Hodge

Introduction

An evaluation of the safety of fluoride prophylaxis in dental health care must be based on a knowledge of the injurious effects of excess fluoride intake. Therefore, the topic first discussed gives a brief review centered around the summary of toxic effects (Table 1).

The hazards attending the use of fluoride tablets or drops have been separated into two groups: a) those of primary concern-dental fluorosis; systemic adverse reactions, and b) those of secondary concern-acute poisoning; allergy; repeated overdose for brief periods. These topics will be discussed in some detail.

Fluoride Toxicology

Among the many effects of fluoride (real or purported) are a few that have been so well studied that quantitative dose-effect relations (Table 1) can be estimated albeit with variable numerical certainty (Hodge and Smith, 1965). Only three of these effects have been observed in man: acute poisoning - death following a single dose; chronic poisoning - crippling fluorosis, the final stage of advanced osteofluorosis, first detectable as an increase in the radio-graphic density of the skeleton (osteosclerosis); and dental fluorosis. Five other chronic fluoride effects have been well studied in experimental animals: kidney injury, anemia, interference with reproduction, changes in thyroid structure or function, and body weight loss. These effects are seen only after exposures of months or years at or near the concentrations listed. The species named in Table 1 are relatively susceptible; other species may re-quire higher concentrations or longer exposures (never

Table I. SELECTED TOXIC EFFECTS OF SOLUBLE INORGANIC FLUORIDES GIVEN BY MOUTH.

ACUTE POISONING

	IN MAN	IN ANIMALS
Certainly lethal dose	± 50 mg F/kg	± 50 mg F/kg
for a 70 kg individual	2500 - 5000 mg F	
for a 10 kg child	535 mg F	

CHRONIC POISONING

IN ANIMALS	ppm F *
Kidney injury (rat) - in water, months.	100
Anemia (rat) - in water, months.	100
Reproduction interference (cattle) - in feed, years	60
Thyroid disturbance (many species) - in feed or water, years	50
Body weight loss (dairy cow) - in feed, years	40

	IN MAN	IN ANIMALS	ppm F *
Crippling fluorosis absorbed for years	10 -> 25 mg F/day	Lameness (cattle)	> 50
Dental fluorosis in drinking water for the first 5-8 years of life	> 2 ppm F **	Dental mottling (cattle)	< 20

* Minimal concentrations in feed or water
** For cosmetic damage : extensive white areas or brown stain

confirmed in man).

Time is an important factor in the toxic effects of fluoride. Lethal doses in man often cause death within 2-4 hours (Note that Table 1 gives estimates of <u>certainly</u> lethal doses not <u>minimal</u> lethal doses.). On the assumption that the lethal doses for children would be proportional to body weight (an assumption not supported by firm data), a certainly lethal dose is suggested for a 10 kg child corresponding approximately to a one year old. The acute lethal doses of fluoride for some species of animals are in the same order of magnitude as the certainly lethal dose for the human adult.

Chronic effects follow protracted exposures. Crippling fluorosis as an occupational disease follows exposures estimated at 10 to over 25 mg of fluoride daily during periods of 10-20 years. The pattern of crippling fluorosis comprises hypermineralization (osteosclerosis), calcification of ligaments, and exostoses. In animals, osteoporosis and osteomalacia have also been observed in severely affected bone. In certain parts of the world, crippling fluorosis develops in individuals drinking water containing elevated amounts of fluoride, for example, 10 or more ppm; this was <u>not</u> seen in communities in the United States where the water supplies contained up to 20 ppm. About 10% of the lifetime residents of a U.S. community where the drinking water contained 8 ppm F exhibited detectable osteosclerosis. Osteosclerosis never occurs in exposed workmen in the U.S. whose urinary fluoride excretion is maintained below 5 mg F per liter (Irwin, quoted by Hodge and Smith, 1970). When urinary fluorides equal or exceed 9 mg F per liter, several years are required for the development of detectable osteosclerosis; the pelvic vertebrae show the earliest X-ray evidence of increased mineralization (Hodge and Smith, 1977).

Dental fluorosis of moderate (extensive white areas or brown stain) or severe (pitted, grooved, stained surfaces) degree may occur in the permanent teeth of children who drink water containing over 2 (to 8 or more) ppm of fluoride during the first 5-8 years of life, i.e., when the ameloblasts are forming enamel. From the classical epidemiologic studies of Trendley Dean, the water concentrations in American communities have been mathematically related to prevalence and severity of dental fluorosis. The mechanism by which fluoride impairs the function of ameloblasts is unknown. Dean (1942) showed that in a given community the severity of fluorosis was not uniform but where the water contained 1 ppm fluoride, 10-20% of

children showed none to mild fluorosis. Where community
water supplies contained naturally occurring fluorides at
concentrations of 4 ppm or more, children 12 to 14 years
of age displayed none to very mild or mild fluorosis together with an increasing prevalence of moderate and
severe fluorosis, as fluoride concentration increased.
The basis for the variation among individuals has not been
determined but differences in water consumption may
account for much of it. One important fact often over-
looked is the "normal" incidence of enamel metaplasia or
hypoplasia; one in five children in any community with
a low water fluoride concentration will show evidence of
non-fluoride tiny white spots or "milky", opalescent
areas. Children in a community in a temperate climate
in which the drinking water contains 1 ppm F show about
the same frequency and severity of the nonfluoride opa-
cities.

The most important and widely disregarded fact about
dental fluorosis is this: "no safe established daily intake
exists", i. e., the maximal amount in mg fluoride which
consumed daily does NOT produce cosmetically damaging
extensive white areas or brown stain in some individuals
has not been fixed. From the numerous studies of fluoride supplements particularly of tablet administration, a
fair approximation of a safe daily dose is emerging; this
fact will be discussed later. In contrast, the drinking
water fluoride concentration below which moderate fluorosis does not occur is established with a remarkable
quantitative certainty. In the United States, no moderate
fluorosis develops in children living in a temperate
climate where the drinking water concentration is 2 ppm
of fluoride or less (Galagen and Lamson, 1953).

Hazards of Primary Concern

Dental Fluorosis

A substantial number of clinical studies attest to the
value of systemic fluoride supplements, especially in tablet
form, in the prevention of dental caries. Few of the investigators mention dental fluorosis although some observations have been recorded (e. g., those of Arnold, McClure
and White, 1960; Feltman and Kosel, 1961; Ericsson and
Ribelius, 1971; Aasenden and Peebles, 1974; Margolis
et al., 1975; Hennon et al., 1976; Driscoll et al., 1976).
The early records gave valuable guidance but were limited.
Arnold et al. examined 32 children at NIDR after about

10 years on their regimen; 4 exhibited dental fluorosis, 3 of questionable and 1 of very mild degree (presumably of the permanent dentition). Feltman and Kosel initially gave 1 mg F per day from infancy and later "eliminated" the fluorosis that occurred by reducing the dose to 0.5 mg daily. Ericsson and Ribelius concentrated their attention on whether breast-feeding during the first year of life would show a difference in fluorosis as compared with dry milk formula feeding. The teeth were examined in 8 to 9 year olds in Uppsala (water 1.2 ppm F) and in all primary grade children (only 25 of them) in Billesholm (water 5.5 ppm F). Predictably, dental fluorosis was more severe in Billesholm, however, there was little difference in the incidence and severity of mottling whether breast-fed or formula-fed in either community (also little difference in caries rates). Margolis et al. saw "none of the mottling effect... in the deciduous teeth" of children given 0.5 mg F as fluoride-vitamin drops beginning at ages one to four months and whose dose was increased to 1 mg fluoride daily from the third through the tenth years. The appearance of the permanent teeth was not mentioned; re-examination of these children after the eruption of canines and premolars would give valuable information. Hennon et al. placed one group of infants 1 to 14 months of age on a fluoride regimen: 0.5 mg fluoride with vitamins up to age 2 and then 1 mg F daily until the study was terminated at age 7. The few erupted permanent teeth exhibited increased fluorosis but not of a "cosmetically unacceptable" degree. Later examinations would add appreciably to the value of this study also.

Aasenden and Peebles provided the first group indices of fluorosis in terms of distribution of severity as classified by Moller's (1965) modification of Dean's standard for clinical diagnosis (although abnormal wear was not mentioned). Three groups of children from the Boston suburbs were formed. Group I, about 100 children residing in non-fluoridated communities, received fluoride supplements "continuously for 4 or more days a week since the age of 1 week to 4 months". Daily doses were 0.5 mg F "mostly as a liquid NaF-vitamin preparation" before the age of 3 years and 1 mg F thereafter "as NaF tablets or drops with or without vitamins" up to age 12. Group II, 93 children, was untreated (i.e., for less than 30 days during early childhood. Group III, 92 children, resided in fluoridated communities (1 ppm F). The percentage distributions are given in Table 2A; selected data from Dean's epidemiological studies are added in part B.

Table 2. Severity of Dental Fluorosis in Groups with Different Fluoride Intakes

	Number	Group Index of Fluorosis	Percentage Distribution*					
			N	Q	VM	M	MO	S
A. Aasenden & Peebles								
Gp. I, F supplement	100	0.88	16	17	34	19	14	--
II, Control	93	0.07	83	13	3	1	--	--
III, F water – 1 ppm	92	0.40	37	30	22	9	2	--
B. Dean								
Clovis, N.M. 2.2 ppm		1.4	13	16	24	3.5	11	.7
Lima, Ohio 0.3 ppm		0.09	84	14	2	--	--	--
Joliet, Ill. 1.3 ppm		0.46	40	34	22	3	--	--

*The degrees of severity of fluorosis are those established by Dean: M–mild, N–normal, Q–questionable, VM–very mild, MO–moderate, S–severe.

Plotting the percentages against the degree of effect, Fig. 1, (compare Hodge and Smith, 1965, p. 455) shows that a) the curve for the control group II coincides nearly exactly with the distribution marked "L" for Lima (0.3 ppm F), b) the curve for group III, comprising children from fluoridated Boston suburbs, and the distribution marked "J" for Joliet (1.3 ppm) correspond closely except for slightly higher percentages of mild and moderate fluorosis in group II, and c) the curve for group I which received supplemental fluoride and the distribution marked "C" for Clovis (2.2 ppm) are similar in percentages of normal and questionable degrees, group I has a higher percentage of very mild, a lower percentage of mild, similar percentages of moderate, and none severely fluorosed. By entering a graph of the community index of severity vs. the logarithm of fluoride concentration (Hodge, 1950), the group I index of 0.88 interpolates to an equivalent of about 1.8 ppm F in naturally fluoridated communities (Fig. 2). The validity of such a comparison depends on the ability of Aasenden to classify each tooth quantitatively as Dean would have. Systematic differences might have been anticipated, e.g., Ericsson and Ribelius recently acknowledged differences between the grading of severity by two clinical colleagues. However, the remarkable conformity of Aasenden and Peebles' data (curves in Fig. 1) to Dean's data (initials in Fig. 1) gives numerical evidence of the concordance of the diagnosis by Aasenden. Equating the effects of the supplement regimen to those of a fluoridated drinking water supply containing more than 1 ppm F may be acceptable. In the first place, a greater reduction in dental caries was observed by Aasenden and Peebles in the supplement group: approximately half as many decayed and filled surfaces occurred in group I as in group III; 37% of the group I children were caries-free, compared with 10% of the children in III. In the second place, Aasenden and Peebles were struck by the undesirable appearance of a few (four or five) cases "although there was no discolouring or pitting of the enamel". Dean (1942) had described a community index of 0.4 to 0.6 as "borderline" in the sense that when "the index rises above 0.6 it begins to constitute a public health problem warranting increasing consideration." Aasenden and Peebles agreed and concluded "that the doses of F given during the first years of life in this study are at the very borderline of the tolerable limit." Infante

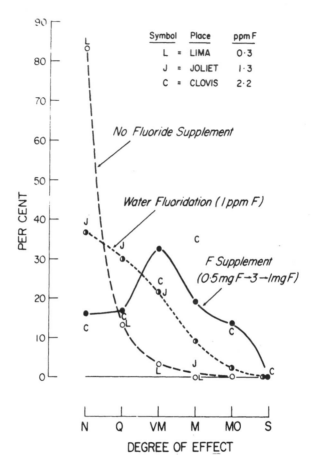

Figure 1. Distribution of dental fluorosis severity in groups with different fluoride intake. The <u>curves</u> display the percentages of each group in Aasenden and Peebles' study in the categories of dental fluorosis originally specified by Dean. Long dashed line: O-group II; short dashed line: ⦿-group III; solid line: ●-group I. The initials, L, J and C, designate Dean's percentages for Lima, Ohio; Joliet, Illinois; and Clovis, New Mexico, respectively.

(1976) on theoretical grounds estimated that for a six year old who had been on the fluoride supplement regimen since birth, the fluoride intake would roughly correspond to an index of 1.4 which interpolates (Fig. 2) to match a naturally fluoridated water supply of 2.4 ppm. Photographic evidence of "chalky white" streaked enamel was published by Grossman (1975) who commented "that children who actually receive the full recommended medicinal fluoride dose* regularly throughout childhood are at a greater risk of developing dental fluorosis than children drinking fluoridated water."

During the past five years, an unfortunate misunderstanding developed which may lead to an unacceptable degree of dental fluorosis in some children. For perspective, a bit of history will be restated. (1) Dean showed that dental fluorosis (originally called mottled enamel) increased in prevalence and degree with increasing fluoride concentration naturally in drinking water. (2) Dean later showed that dental caries prevalence and extent decreased with increasing fluoride concentration in drinking water. At 1 ppm F, caries reductions of 50 to 70% accompanied cosmetically "normal-appearing" teeth (fluorosis index of 0.6 or less). (3) When fluoride was added to a community water supply in a temperate climate to bring the fluoride concentration up to about 1 ppm, caries incidence was reduced without increasing the number of children with moderate or severe mottled enamel, i.e., the practice was effective and safe as regards appearance. (4) In an effort to bring the dental health benefits to children who had no access to community water systems, various ways of supplying fluoride orally were proposed and several were tried, including regimens of fluoride tablets, capsules, or pills (e.g., among the earlier studies were those of Harootian, 1943; Strean and Beaudet, 1945; Larson, 1947; Stones et al., 1949; Held and Piquet, 1954; Held, 1955; Bibby et al., 1955; Wrzodek, 1959; Arnold et al., 1960; Feltman and Kosel, 1961). Fluoride was supplied as bone flour, calcium fluoride, sodium fluoride or sodium monofluorophosphate in daily amounts that at the start usually corresponded more or less closely to the amount of fluoride (1 mg) which would be obtained from drinking a liter of water containing 1 ppm F (a widely accepted rule-of-thumb for the usual consumption of adults in a temperate climate).

The importance of McClure's estimates of the fluoride intake by children who drink water containing 1 ppm F

*(0.5 mg F daily to age 3, 1 mg daily thereafter)

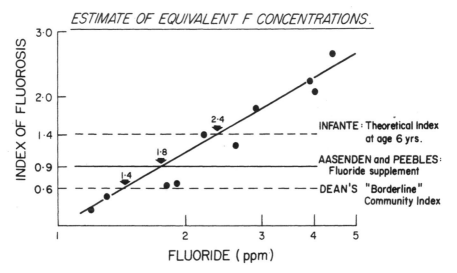

Figure 2. Conversion of tablet schedule to equivalent
fluoride concentration in drinking water by community
fluorosis index. A linear relation is shown when the community indices of dental fluorosis (Dean's data) are plotted
(●) against the logarithms of the fluoride concentrations in
each community water supply. The arrows indicate the
fluoride concentrations (1.4, 1.8 and 2.4 ppm) corresponding
to the observations listed at the right.

justifies a brief resume of his thinking (McClure, 1943). He began with Adolph's 1933 estimate of "the daily water requirement as equal to 1 cc. per calorie of energy in the daily diet." Water consumption by children is not only variable but "the requirement of water is largely met by preformed water in the food or by liquid food, particularly milk." McClure assumed that drinking water provided a) 25% or b) 33% of the requirement plus c) 10% or d) 20% of the food-borne water (i.e., added during cooking) "as consisting of water of drinking water origin." The total fluoride daily intakes from drinking water containing 1 ppm F were estimated to range from 0.39 to 0.56 mg for 1 to 3 year olds, from 0.52 to 0.74 mg for 4 to 6 year olds, from 0.65-0.93 mg for 7 to 9 year olds, and from 0.81 to 1.16 mg for 10 to 12 year olds.

The impact of McClure's estimates was unmistakable. Several dosage plans were tried; nearly all were successful in reducing the incidence of dental caries. A few of these plans will be mentioned.

Because 1 ppm F was held to engender no ill effects, Held estimated the optimal total intake of fluoride for caries prophylaxis and since the average Swiss diet contained 0.38 mg, the fluoride "deficit" (Table 3) could be calculated on the basis of McClure's figures for the intake by children of various ages drinking waters bearing only a trace of natural fluoride (0.1-0.2 ppm F).

Table 3. Held's Proposal for Doses to Supply the Deficit

Ages	No. of 0.25 mg F tablets daily	Total mg F daily
1-3	2	0.5
4-6	2-3	0.5-0.75
7-9	3-4	0.75-1.0
10-12	4-5	1.0-1.25

The fluoride-tablet regimen followed in the 1960 study of Arnold, McClure and White, based understandably on McClure's earlier estimates, "differed in some important ways from those followed in previous studies." The majority of the children entered this program before the age of 6, some soon after birth. The investigators estimated the total daily fluoride intake of children drinking water containing 1 ppm F as 0.2-0.3 mg F from the average diet plus 0.4-0.6 mg F from water for the average child less than 3 years old (Note that McClure's 1943 estimate was for children 1 to 3 years old). The instructions Arnold et al. gave parents for the use of tablets each containing 1 mg F were:

Children 0-2 years old: Dissolve 1 tablet in a quart
                           of water for making formula
                           or for drinking.
Children 2-3 years old: Give 1 tablet every other day.
Children 3-10 years old: Give 1 tablet daily.

Caries incidence was reduced and in a total of 32 children
examined, fluorosis did not exceed a mild degree in the four
children affected.   This regimen was not accepted in its
entirety perhaps because making a solution for infants
proved to be bothersome. On the other hand, the recommended dose for children 3 years of age and older represents a widely held opinion today although other plans
deferring the start of 1 mg daily doses to ages 6 to 7 are
offered (Hotz, 1974; Stephen Wei, personal communication,
1977; Fomon and Wei, 1976).*

No such agreement exists on the optimal doses for
children from birth to age 3. One plan tested in the 1960's
utilized commercially available vitamin-fluoride drops and
tablets. Subjects between the ages of 0 and 24 months
received drops containing 0.5 mg F, those two years old
and older a chewable tablet containing 1 mg F daily (Hennon
et al., 1966). In 1970, McClure wrote, "It is generally
recognized now that tablets containing 0.5 mg of fluoride
may be started shortly after 6 months and continue to 3
years, at which time a 1 mg tablet should be ingested daily
until the child is 8 or 10 years old." A schedule calling
for 0.5 mg F daily in the first 3 years and 1 mg F daily
therafter has been recommended by certain manufacturers
of fluoride tablets and in 1972 by the Committee on Nutrition of the American Academy of Pediatrics (Filer et al.,
1972).

The "unfortunate misunderstanding" refers to the emer-
gence of a concept that the recommended daily intake of
0.5 mg F from birth to age 3 represents the optimal
intake and therefore that a fluoride deficiency exists if
a child ingests daily less than 0.5 mg F. A recent finding
in a fluoridated area that 68% of the children less than
2 years old were receiving less than 0.5 mg F, led to an

---

*A remarkably successful dental health regimen of fluoride
tablet administration in Norway (Per Lokken, personal
communication, 1977) recommends the following doses:

Dose	Age
0.25 mg daily	< 1 year
0.5	1-6
0.75	6-12
1.0	> 12

opinion -- "Fluoridated water supplies: An inadequate source of fluoride for children" -- that presumably may not be limited to the investigators (Schwab and Schwartz, 1975) who concluded that "The administration of fluoride to all children during the years of tooth formation, including those previously thought to have been protected by addition of fluoride to the water supply, must therefore be recommended to insure optimal prophylaxis against the formation of dental caries." The origin of the case Dean made for water fluoridation had been mistakenly replaced by the estimate of what might constitute an equivalent fluoride supplement. Following Grossman's warning, Schwartz and Schwab (1975) reconsidered and agreed that "the overzealous physician may possibly contribute to an increased incidence of tooth mottling."

The authoritative Accepted Dental Therapeutics (36th edition, 1975), recommends another dosage schedule with lower doses from birth to 3 years of age: taking into account the fluoride concentration in the water supply, 0.25 mg F/day from birth to 2 years and 0.5 mg F/day from age 2 to 3 years. The paucity of quantitative information relating the index of fluorosis to the systemic fluoride regimen together with the widespread (and growing) use of tablets and drops or other means of F supplementation by mouth argues strongly for the initiation of long-range, rigorously controlled clinical studies embracing the most promising of several alternate modes and daily dose plans. Lacking an intensive attack, the optimal daily allowance for fluoride supplements may remain unclear for many years.

Among the preventive dentistry programs are those in which tablets are given to school children under the direct supervision of the teacher. For children whose enamel formation is practically complete, e.g., in grades 1 to 3 or above, the risk of fluorosis is predictably nearly nil from additional fluoride treatment. No fluorosis was observed (Driscoll, 1974, and personal communication, 1976) in a 55-month study of school children in the first and second grades who received daily from their teacher either (1) placebo tablets, (2) one F tablet (1 mg F), or (3) two F tablets (2 mg F) at least 3 hours apart, to chew, swish and swallow. "Questionable" fluorosis was observed in 13% of the control group, 19% of the 1-tablet group, and 21% of the 2-tablet group. Only 1 child in the 2 tablet group exhibited very mild fluorosis provisionally attributed to the treatment. Marthaler (1967) commented that "after the period during which the teeth mineralize, the body may tolerate elevated doses of fluo-

ride without damage". Several recent reports illustrate
the efforts to explore this possibility (Heifetz and
Horowitz, 1974; Horowitz et al., 1972; Heifetz et al.,
1974). It should be acknowledged, however, that the intent
of elevating the fluoride concentration in school water
was to compensate for the fraction of daily water consumption represented.

Systemic Adverse Reactions

In a sizable number of studies involving thousands of
subjects in which the efficacy of fluoride tablets has been
demonstrated, systemic ill effects are rarely considered
and have been reported only once. "One percent" of
Feltman and Kosel's (1961) subjects, "gravid women and
children of all ages", suffered from "dermatologic, gastrointestinal and neurological" ailments. No documentation
was provided. The causes of these "undesirable side
effects" remain obscure. Fluoride cannot with certainty
be identified as the etiologic agent.

Hazards of Secondary Concern

Acute Toxicity

If the total amount of fluoride available to a child is
large enough, and if an excessive amount is ingested, the
poisoning can be life-threatening. The 36th edition of
Accepted Dental Therapeutics (1975), in the chapter on
fluoride compounds, page 292, states, "It is therefore
recommended that no more than 264 mg of sodium fluoride
be dispensed at one time." This widely used guideline
intended to prevent serious or possibly fatal home poisonings of children originated at a meeting in the mid-fifties
at the Food and Drug Administration, Washington, D.C.,
attended by Dr. Edwin P. Laug and one other scientist
from the Food and Drug Administration, by Drs. Hodge
and Joseph Muhler, consultants for the FDA, and by a
small group from the Procter and Gamble Company, including Dr. A.W. Radike, Mr. Verling Votaw, one of
the Procter and Gamble vice-presidents, and a few others.
The tragic history of accidental deaths from fluoride, used
as a household insecticide, but carelessly handled and
stored, left unguarded, emphasized the dangerously toxic
nature of solid sodium fluoride. Many fatal home accidental poisonings occurred from 1900 to 1940, in fact,
several hundred people died of acute fluoride poisoning
(Hodge and Smith, 1965). The Procter and Gamble Company had evaluated the acute toxicity of fluorides in mice
and rats and from the animal data using a safety factor,

had estimated that the contents of one or two five ounce tubes of fluoride dentifrice would be tolerated if ingested by a 10 kg child (age 1 year) in a single dose. At the meeting, when the question was raised, "How much fluoride can safely be allowed in the home in one container?" Dr. Hodge made a few simple calculations on the blackboard. a) He and Dr. Frank A. Smith had estimated from human fluoride poisonings that 5,000-10,000 mg (av. 7500 mg) of sodium fluoride would be a "certainly lethal dose" for a seventy kg adult. b) Assuming that the acute toxic dose of sodium fluoride varies directly with body weight, the certainly lethal dose for a 10 kg child would be about 1070 mg (535 mg F). c) Black et al. (1949) had given solutions of sodium fluoride to patients by intravenous infusion on several successive days in amounts constituting substantial fractions, e.g., 1/4 or more of the "certainly lethal dose" without producing signs of "acute or chronic fluoride toxicity." d) Mortality in animals decreases rapidly with decreasing dose, e.g., rats survive doses as large as one-half the LD50 (de Lopez et al., 1976). e) For a 10 kg child, one quarter of the estimated certainly lethal dose (1070/4 or 267 mg) was predicted to be non-lethal, and therefore an estimate of the "safe" quantity for home storage. Someone sitting at the table said, "267? That's an odd kind of a number. Why not a good round number?" Someone else proposed, "What about 264 mg?" This seemed to be a satisfactory "round" number, acceptable to all present. A five ounce tube of stannous fluoride dentifrice contained less than the proposed limit of fluoride expressed as sodium fluoride equivalent, and was the largest tube marketed for years. As other fluoride products for home dental health use have appeared, this limit on the total fluoride in one container has been applied. Experience during the past score of years has borne out the safety of the 264 mg limit; to the author's knowledge, no death has been recorded.

The Food and Drug Administration later rounded off the number a little more generously so that at present, the limit is 300 mg of NaF per retail package (¢. 301. 201-10-iv). The American Dental Association Council on Dental Therapeutics has recently re-affirmed the applicability of the 264 mg limit to mouth rinse products that can readily be consumed but "for products in individual dose containers...the Council will limit acceptance to products packaged to contain a maximum of 300 mg of sodium fluoride" (Council on Dental Therapeutics, 1975).

Accidental ingestion of 20 to 100 tablets containing 1 mg F each probably would be followed by signs and

symptoms no more drastic than nausea and vomiting,
perhaps by abdominal pain and prostration (Infante, 1974;
Thienes and Haley, 1972; Gosselin, R. E. et al., 1976).

Allergy?

Margolis et al. (1975) did not "observe any cases that
could be identified as having allergic manifestations attributable to fluoride." No other comments are available.

Repeated Overdose for Brief Periods

Because Dean's epidemiological studies of the relation
between dental fluorosis and community water concentrations included only the teeth of children who had never
been away from their water supply for more than 30 days,
it has been tacitly assumed that long-continued exposures
to elevated fluoride levels were necessary. Two recent
findings challenge this assumption: a) Suttie and Faltin
(1971) showed in cattle that shorter periods of high fluoride
intake were more damaging than longer periods with lower
intake; b) Angmar-Mansson et al. (1976) confirmed
Kruger's observation that in the rat an intraperitoneal
dose of 0.1 mg F/kg body weight is a marginal mottling
dose (for incisor enamel) and that a week's exposure to
20-30 ppm F in drinking water would disturb the normal
mineralization of incisor enamel. In cattle, plasma levels
of about 0.3 ppm F were associated with severe mottling;
in rats, faulty mineralization occurred at plasma F levels
of about 0.2 ppm. As has been long known, man is more
sensitive to this fluoride effect than rats or cattle. Based
on the data from Taves' laboratory (Guy et al., 1976)
relating water fluoride to plasma fluoride ion concentrations and on Dean's community indices of fluorosis, dental
fluorosis of moderate to severe degree should develop
in man when plasma F ion levels reach 0.05 to 0.1 ppm
(Table 4).

These studies do not reverse the long accepted belief
that long residence is required in a community with a
drinking water supply bearing more than 2 ppm fluoride
to engender cosmetic or structural damage. Further in-
quiry, however, is indicated.

Summary

1. The hazard of primary concern in the supplemental
   use of fluoride tablets and drops is dental fluorosis.

Table 4   Drinking water fluoride concentrations related
to plasma fluoride ion concentrations in residents of communities using these waters (Taves' data) and to the community indices of fluorosis (Dean's data).

Water F (ppm)	Plasma F * (ppm)	(uM)	Index of** Fluorosis
0.1	0.008	0.4	0.01
1.0	0.018	1.	0.3
2.1	0.046	2.4	1.4
5.6	0.086	4.5	3.3

*Guy et al., 1976
**Dean, 1942

2. The widely recommended dosage regimen which
    gives infants 0.5 mg F daily for the first two
    years of life and 1 mg F daily from age 3 on is
    probably too high.   Such children are at "greater
    risk of developing dental fluorosis than children
    drinking (optimally) fluoridated water".

3. An alternate regimen recommended by the Council
    on Dental Therapeutics of the American Dental
    Assoc. provides 0.25 mg F daily for the first 2
    years, 0.5 mg F daily from 2 to 3 years and 1 mg
    daily thereafter.   No epidemiological assessment
    has been made of the safety of this dosage.

4. Long-term, controlled clinical studies of one or
    more dosage plans are needed.

5. Tablets and drops are not intended to be used in
    communities with optimally fluoridated water
    supplies.

6. The risk of dental fluorosis in school children
    (age 6 and older) in non-fluoridated communities
    who receive tablets only on school days as partici-
    pants in school programs is practically nil.

7. Systemic adverse reactions to fluoride tablets have
    been reported in only one of the numerous studies.

8. Acute toxic effects of serious proportions are highly improbable from the ingestion of 264 to 300 mg fluoride as a single dose.

Acknowledgement

The author thanks those who have commented on various parts of this paper, especially W. S. Driscoll, S. B. Heifetz, H. S. Horowitz, E. Newbrun, D. R. Taves, and S. H. Y. Wei.

Bibliography

Aasenden, L. and Peebles, T.C. (1974). Effects of fluoride supplementation from birth on human deciduous and permanent teeth. Arch. Oral Biol. 19; 321-326.

Accepted Dental Therapeutics (1975). Council on Dental Therapeutics, American Dental Association, Chicago, 36th ed., 288-305.

Angmur-Mansson, B., Ericsson, Y. and Ekberg, O. (1976). Plasma fluoride and enamel fluorosis. Calcif. Tiss. Res., 22; 77-84.

Arnold, F.A. Jr., McClure, F.J. and White, C.L. (1960). Sodium fluoride tablets for children. Dent. Progr., 1; 8-12.

Bibby, B.G., Wilkins, E. and Witol, E. (1955). A preliminary study of the effects of fluoride lozenges and pills on dental caries. Oral Surg., 8; 213-216.

Black, M.M., Kleiner, I.S. and Bolker, H. (1949). The toxicity of sodium fluoride in man. N.Y. State J. Med., 49; 1187-1188.

Buttner, W., Henschler, D. and Patz, J. (1973). Kariesprophylaxe durch fluorid-einnahme. Dtsch. Med. Wochschr., 98; 751-756.

Council on Dental Therapeutics. (1975). J. Am. Dent. Assn., 91; 1250.

Dean, H. Trendley (1942). The investigation of physiological effects by the epidemiological method. Fluorine and Dental Health, AAAS, ed. Forest Ray Moulton, Washington, D.C., pp. 23-31.

de Lopez, O.H., Smith, F.A. and Hodge, H.C. (1976). Plasma fluoride concentrations in rats acutely poisoned with sodium fluoride. Tox. Appl. Pharmacol. 37; 75-83.

Driscoll, W.S., Heifetz, S.B. and Korts, D.C. (1974). Effect of acidulated phosphate-fluoride chewable tablets in school children: results after 30 months. J. Am. Dent. Assoc., 89; 115-120.

Driscoll, W.S., Heifetz, S.B. and Korts, D.C. et al. (1977). Effect of acidulated phosphate-fluoride chewable tablets in schoolchildren: results after 55 months. J. Am. Dent. Assoc., 94; 537-543.

Ericsson, Y. and Ribelius, U. (1971). Wide variations of fluoride supply to infants and their effect. Caries Res., 5; 78-88.

Feltman, R. and Kosel, G. (1961). Prenatal and postnatal ingestion of fluorides, fourteen years of investigation. Final Report. J. Dent. Med., 16; 190-198.

Filer, L.J. and the Committee on Nutrition, American Academy of Pediatrics. (1972). Fluoride as a nutrient. Ped., 49; 456-460.

Fomon, S.J. and Wei, S.H.Y. (1976). Prevention of Dental Caries, Part II, Chap. 4. In Nutritional Disorders of Children: Prevention, Screening and Follow-up, ed. Fomon, S.J. U.S. Gov't. Printing Press, In press.

Galagan, D.J. and Lamson, G.G. (1953). Climate and Endemic Dental Fluorosis. Pub. Health Reports, 68; 497-508.

Gosselin, R.E., Hodge, H.C., Smith, R.P. et al. (1976). Clinical Toxicology of Commercial Products. IV Edition. Williams and Wilkins, Baltimore.

Grossman, E.R. (1975). Letter, J. Ped.,87; 840-841.

Guy, W.S., Taves, D.R. and Brey, W.S. (1976). Organic Fluorocompounds in Human Plasma, Prevalence and characterization. Am. Chem. Soc. Symposium Series #28, Biochemistry Involving Carbon-Fluorine Bonds, Robert Filler, ed., pp. 117-134.

Harootian, S.G. (1943). The influence of administration of bone flour on dental caries. J. Am. Dent. Assoc.,30; 1396-1399.

Heifetz, S.B., Horowitz, H.S., Driscoll, W.S. (1974). Utilization of fluorides in areas lacking central water supplies. J. Can. Dent. Assoc.,40; 136-146.

Heifetz, S.B. and Horowitz, H.S. (1974). Effects of school water fluoridation on dental caries: interim result in Seagrove, N.C., after four years. J. Am. Dent. Assoc., 88; 352-355.

Held, A.J. and Piquet, F. (1954). Prophylaxis of dental caries by fluoridated tablets. First results. Schweiz. Mschr. Zahn., 64; 694-697.

Held, A.J. (1955). Experiences and problems of water fluoridation. Dtsch. Zahnaerztl. Z., 10; 271-278.

Hennon, D.K., Stookey, G.K. and Beiswanger, B.B. (1966). Fluoride-vitamins: Effects on caries and fluorosis in sub-optimal fluoride areas. J. Dent. Res., 55; Abstract 90.

Hodge, H.C. (1950). The concentration of fluorides in drinking water to give the point of minimum caries with maximum safety. J. Am. Dent. Assoc., 40; 436-439.

Hodge, H.C. and Smith, F.A. (1965). Fluorine chemistry. ed., J. H. Simons. Vol. IV, Academic Press, N.Y.

Hodge, H.C. and Smith, F.A. (1970). Air quality criteria for the effects of fluorides on man. J. Air Pollut. Control Assoc., 20; 226-232.

Hodge, H.C. and Smith, F.A. (1977). Occupational fluoride exposure. J. Occup. Med., 19; 12-39.

Horowitz, H.S., Heifetz, S.B. and Law, F.E. (1972). Effects of school water fluoridation on dental caries: final result in Elk Lake, Pa., after 12 years. J. Am. Dent. Assoc., 84; 832-838.

Hotz, P. (1974). Discussion, International Workshop on Fluorides and Dental Caries Reductions, University of Maryland.

Infante, P.F. (1974). Acute fluoride poisoning - North Carolina. J. Pub. Health Dent., 34; 281-282.

Infante, P.F. (1975). Dietary fluoride intake from supplements and communal water supplies. Am. J. Dis. Child., 129; 835-837.

Margolis, F.H., Reames, M.R., Freshman, E. et al. (1975). Fluoride - Ten year prospective study of deciduous and permanent dentition. Am. J. Dis. Child.,129; 794-800.

Marthaler, T.M. (1967). The values in caries prevention of other methods of increasing fluoride ingestion, apart from fluoridated water. Intnl. Dent. J., 17; 606-618.

McClure, F.J. (1943). Ingestion of fluoride and dental caries. Quantitative relations based on food and water requirements of children 1 to 12 years old. Am. J. Dis. Child., 66; 362-369.

McClure, F.J. (1970). Alternative uses of fluoride. In Water Fluoridation, The Search and The Victory, USDHEW, NIH, Nat'l. Inst. Dent. Res., Bethesda, Md., p. 176.

Møller, I.J. (1965). Dental Fluorose og Caries, Rhodos, Copenhagen.

Schwab, J.G. and Schwartz, A.D. (1975). Fluoridated water supplies: An inadequate source of fluoride for children. J. Pediatr., 86, 735-736; 87, 841.

Stones, H.H., Lawton, F.E., Bransby, E.R. et al. (1949). The effect of topical applications of potassium fluoride and of the ingestion of tablets containing sodium fluoride on the incidence of dental caries. Br. Dent. J., 86; 263–271.

Strean, L.P. and Beaudet, J.P. (1945). Inhibition of dental caries by ingestion of fluoride-vitamin tablets. N.Y. State J. Med., 45; No. 20.

Suttie, J.W. and Faltin, E.C. (1971). Effect of a short period of fluoride ingestion on dental fluorosis in cattle. Am. J. Vet. Res., 32; 217–222.

Thienes, C.H. and Haley, T.J. (1972). <u>Clinical Toxicology</u>. 5th ed., Lea and Febiger, Philadelphia. pp. 176–179.

Wrzodek, G. (1959). Does the prevention of caries by means of fluorine tablets promise success? Zahnaerztl. Mitt., 47; 258–262. U.S. Dept. of Commerce translation No. 59-18893.

12

Fluoridation and Bone Disease in Renal Patients

William J. Johnson, Donald R. Taves and Jenifer Jowsey

The question of whether fluoridation contributes to bone disease in patients with renal disease is an important one that deserves attention and further study. It is logical to be concerned about this possibility because of the important role of the kidney in the elimination of fluoride. However, the subject is extremely complicated for several reasons. Ever since the initial description of bony changes induced by fluoride in persons with normal renal function, there has been continuing controversy regarding whether these effects are harmful or beneficial (Johnson, 1965; Hodge and Smith, 1968; Jowsey, Riggs, and Kelly, 1978). Renal disease in itself produces additional complexities because of the effects of retained phosphate, depressed ionized calcium, excessive parathyroid hormone in the circulation, impaired formation of 1,25-dihydroxy D_3 by the diseased kidney, and impaired collagen synthesis, all of which adversely affect the integrity of the skeleton. The added significance of fluoride in this setting has been difficult to interpret (Rao and Friedman, 1975).

In the United States, there have been no reported cases of skeletal fluorosis in persons who drink water containing only one part per million (ppm) of fluoride. However, since no systematic studies have been carried out in patients with renal insufficiency, this possibility cannot be excluded with certainty. If it could be shown that no adverse effect occurred when the total intake of fluoride by patients with renal disease is higher than would be obtained by drinking fluoridated water with 1 ppm of fluoride, then one could assume that no adverse effect was occurring in the population.

There are two obvious settings in which patients with renal disease will receive more fluoride than would occur in normal subjects who drink fluoridated water. One of these situations occurs when patients with renal disease drink water containing more than 1 ppm of fluoride. The other situation is in patients maintained by intermittent hemodialysis when the dialysate is made up with fluoridated water. Hemodialysis under these circumstances introduces fluoride because the concentration in the dialysate (1 ppm equals 50 µM) is much higher than in the blood, and the fluoride ion moves from dialysate to patient.

Fluoride Metabolism

Ingested fluoride is absorbed mainly in the upper gastrointestinal tract (Hein et al., 1956; Hodge and Smith, 1968), and trace doses of fluoride leave the blood in minutes, concentrating in bone and kidney (Wallace-Durbin, 1954; Hein et al., 1956). Most of the orally administered fluoride is excreted in the urine under normal circumstances (Hein et al., 1956; Carlson, Armstrong, and Singer, 1960). Fluoride steadily accumulates in the human skeleton during most of a lifetime, reaching a maximal concentration at about the sixth decade (Hodge and Smith, 1968).

Although the daily urinary content of fluoride remains normal in patients with renal insufficiency, as reported by Smith and associates (1955), this does not imply that excretion is normal. The renal clearance of fluoride is directly related to renal function as measured by the creatinine clearance and the ability to excrete fluoride decreases markedly as renal function deteriorates (Berman and Taves, 1973). Parsons and co-workers (1975) concluded from the limited information available that, in patients with renal failure, fluoride retention was a late event; but little information is available regarding deposits of fluoride in the skeleton during the early stages of renal insufficiency. Once renal function has deteriorated to extremely low levels, serum fluoride concentrations do increase. Retained fluoride is deposited in a nonuniform manner through exchange with hydroxy ions in the apatite mineral of bone (Young, Van der Lugt, and Elliott, 1969). More fluoride accumulates in spongy bone than in compact bone (Kim et al., 1970). The fasting serum fluoride concentration in µM units multiplied by 1,000 is an approximation of the concentration that would be expected in bone as ppm, ashed-weight basis, (Taves and Guy, This monograph). In patients with renal disease, however, this relationship does not hold unless only the forming bone is considered, because the serum levels would be expected to increase markedly, as a result of the disease, over a period of months to years rather than over a lifetime.

Patients are now generally maintained by hemodialysis when renal function deteriorates below 10% of normal. If the

water used to prepare the dialysate contains fluoride,
further accumulation of this ion occurs during dialysis
(Taves et al., 1968). Although such patients may excrete
slightly higher quantities of fluoride in the stool, they are
virtually deprived of renal function and are unable to
excrete fluoride by the usual route. Several investigators
(Taves et al., 1968; Prosser et al., 1970; Nielsen et al.,
1973) have shown that the clearance of fluoride by conventional artificial kidneys closely approximates the clearance
of creatinine, but in the opposite direction. In Nielsen and
associates' patients (1973), concentrations of arterial blood
fluoride were 9 µM before the dialysis (approximately 9 times
above normal) and increased to 19 µM when a dialysate
containing 53 µM of fluoride was used (1 ppm). In patients
with normal renal function, the daily intake of fluoride is 1
to 3 mg, and virtually all of this is excreted by the kidney,
whereas in patients dialyzed with fluoridated dialysate, the
uptake during a single dialysis is 14 to 36 mg, with no
mechanism for elimination from the body. Obviously, during
the course of years, a patient being dialyzed thrice weekly
would accumulate a large amount of fluoride.

The retained fluoride accumulates almost entirely in
bone. In the studies of Nielsen et al. (1973), the disap-
pearance curves of radioactive fluoride (^{18}F) administered
during dialysis or intravenously between dialyses were linear
with respect to time and had similar slopes with both
techniques of administration. These linear plots were
interpreted as indicative of a single-compartment system
which reflected a simple exchange of serum fluoride for bone
hydroxy ions or bone fluoride, governed to a large extent by
the rate of blood flow through bone. In fact, ^{18}F is so
completely cleared by bone in a single passage that it has
been used to measure the blood flow rates through bone
(Wootton, 1974). When patients previously exposed to intravenously administered ^{18}F were dialyzed using fluoride-free
dialysate, the egress of fluoride was much less than the
quantity absorbed when fluoridated dialysate was used--thus
suggesting that, once fluoride enters the skeleton, it
exchanges to a very limited extent. Bone scans confirmed
that ^{18}F is incorporated into the skeleton without detectable
uptake in soft tissues (Hosking and Chamberlain, 1972;
Nielsen et al., 1973). Similar findings have been reported
in animal studies (Wootton, 1974).

Patients dialyzed for more than a year with fluoridated
dialysate show a gradual increase in the serum fluoride
concentration (Siddiqui et al., 1970; Posen, Marier, and
Jaworski, 1971; Jowsey et al., 1972; Cordy et al., 1974),
whereas patients dialyzed using fluoride-free dialysate have

serum levels that remain approximately the same as when they entered long-term dialysis therapy (Table 1). The bone content of fluoride is also greater in patients exposed to fluoridated dialysate than in those exposed to fluoride-free dialysate (Jowsey et al., 1972; Cordy et al., 1974). Whereas these findings are generally accepted, the specific effect of retained skeletal fluoride has been difficult to determine. Before proceeding with a description of patients with renal failure who were exposed to high concentrations of fluoride, it is appropriate to briefly describe the bone disease seen in patients with low fluoride levels.

Table 1. Serum fluoride concentrations.

Patient status	Concentration	
	µg/ml	µM
Normal (adults)	0.02	1.0
Chronic renal failure (Siddiqui et al., 1970)		
Serum creatinine <3 mg/dl	0.05	2.6
Serum creatinine >3 mg/dl	0.09	4.7
Hemodialysis (1 yr or more)		
Without fluoride (Jowsey et al., 1972; Cordy et al., 1974)	0.06	2.8 to 3.4
With fluoride	0.24	12.3 to 13.5

Bone Disease in Patients
With Low Fluoride Levels

The changes in bone seen in patients with chronic renal insufficiency are complex in that there is a spectrum of findings. At one extreme is pure osteotis fibrosis, which shows a pattern of excessive bone resorption. This is seen histologically as excessive numbers of osteoclastic lacunae and reabsorption of bone by osteocytes. Roentgenographically, the bone has a moth-eaten appearance along subperiosteal surfaces. These changes are most easily seen in roentgenograms of the hands and of the distal portion of the clavicle.

At the other extreme is osteomalacic bone disease that is due to a deficiency of the active vitamin D metabolites normally produced by the kidneys. Osteomalacia is recognized histologically by excessive amounts of osteoid (that is, unmineralized bone matrix) and an increase in the fraction of the surface of bone that is forming bone matrix. Roentgenographically, the most easily recognized changes are narrow lines containing very little mineral (Looser zones). These lines probably represent actual fractures because they are frequently followed by callus formation similar to that seen in typical fractures. These two types of bone disease are,

of course, also seen in patients without renal disease.
Patients with chronic renal disease often have a little of
both types.

In addition, the patients occasionally have sclerosis.
Histologically, the sclerosed bone exhibits an increase in
the thickness of the trabeculae, and roentgenographically,
there is a coarsening of the trabecular pattern with
increased opacity of the mineralized tissues. Sclerosis and
increased amounts of osteoid are also features of fluorosis.
Therefore, Kaye and co-workers (1960) considered the possibility that these changes in patients with renal disease were
due to fluoride exposure. They concluded, however, that this
was not so because the bone fluoride concentration in their
cases was considerably lower than that in reported cases of
fluorosis. If this finding is correct and not just the
result of failure to sample recently formed bone, then renal
disease and fluoride cause similar changes. This overlap
makes it very difficult to assess the effect of fluoride per
se in these patients. Very large amounts of new periosteal
bone and calcification of the interosseous ligaments are the
only known features distinctive of fluoride, and these have
not been noted in patients on hemodialysis (Johnson, 1965;
Jolly, Singh, and Mathur, 1969).

Case Reports From High Fluoride Areas

Effects Prior to Dialysis

One case of symptomatic skeletal fluorosis (radiculomyelopathy) has been reported from an area in Texas with natural
fluoride at 2.3-3.5 ppm in the water (Sauerbrunn et al.,1965).
There have been two cases of suspected skeletal fluorosis
(based on X-ray evidence) in the United States with fluoride
at 2-3 ppm in the drinking water (Juncos and Donadio, 1972).
The Department of Nephrology at the Mayo Clinic examines
approximately 100 new patients with end-stage renal disease
each year. Some of these patients reside in areas where the
naturally occurring fluoride concentration in tap water
exceeds 1 ppm. During the course of several years, six
patients have been seen in whom fluoride may have been the
cause of detectable clinical and roentgenographic effects.
Two of these cases have been reported previously from our
institution (Juncos and Donadio, 1972). One patient had
chronic glomerulonephritis, and the others had congenital
renal disease of more than 15 years' duration before skeletal
symptoms developed (Table 2). Most of the patients had high
urine volumes (>3 per day), the fluid being replaced by
copious intake of water or in one instance, tea.

Table 2.   Causes of renal failure in six patients
exposed to high fluoride before dialysis.

Cause	No.
Congenital defects of bladder, ureters	3
Fanconi syndrome	1
Bilateral polycystic kidneys	1
Chronic glomerulonephritis	1

The most distinctive features suggesting fluorosis were
the roentgenographic appearance of the skeleton and the severity of dental mottling (Table 3). The most characteristic
roentgenographic finding was a diffuse increase in bone
density which, in younger patients, assumed a ground-glass
appearance; whereas in older patients, it showed a coarse
trabecular pattern that became more obvious as the skeleton
became more demineralized over the years and with progression
of renal failure. In addition to the increase in bone density and the alteration of trabecular pattern, there was promi-
nent new subperiosteal bone formation, especially in the long
bones of the upper and lower extremities, and calcification
of the interosseous ligaments between tibia and fibula and
between radius and ulna, as well as of the sacrotuberous ligaments of the pelvis and the longitudinal ligaments of the
spinal column in some of the patients. In addition, three patients had pseudofractures, a common feature of osteomalacia.

Table 3.   Roentgenographic findings in six patients
with renal failure who were exposed to high fluoride*
before dialysis.

Finding	Patients
Increased bone density	6
Dental mottling	2
Calcified interosseus ligaments	2
Subperiosteal bone	2
Subperiosteal resorption	0
Fractures	3
Pseudofractures	3

*1.7 to 2.0 ppm.

Despite having severe symptomatic bone disease, none of
the patients showed striking features of hyperparathyroidism,
such as subperiosteal resorption or bone cysts, which, in the
United States, are the more common manifestations of renal
osteodystrophy. Plasma parathyroid hormone concentrations,
although elevated in all of the patients, were relatively low
considering the severity of the bone disease.

In addition, these patients developed severe skeletal
changes or bone pain early in the course of renal failure
when creatinine values were approximately 3 mg/dL. Symptoms

referable to the skeleton varied. Two of six patients were
asymptomatic; four complained of arthralgia, especially of
the knees, and of bone pain on weight-bearing involving the
lower extremities; three of the patients had spontaneous
fractures of metatarsals, ribs, and hip.

Bone biopsy specimens available from four patients
showed a marked increase in the ratio of fluoride to calcium
(Table 4). Biopsy specimens studied by quantitative microradiographic techniques showed a large percentage of bone
surface covered by osteoid as well as thick osteoid seams
with variable degrees of bone resorption and large areas of
new bone formation.

Table 4. Patients exposed to fluoride prior to dialysis.

	Fluoride			Osteoid on bone biopsy	
Case	Water (ppm)	Serum (μM)	Bone* F/Ca	Surface (%)	Width (μM)†
1	1.9	14.1	5.9	65.3	42 ± 2.9
2	2.0	10.1	5.4	46.7	28.8 ± 3.8
3	1.7	5.0	3.5	45.2	21.9 ± 0.8
4	1.7	12.0	3.0	19.4	22.4 ± 2.6
Mean	1.83	10.3	4.4	44.2 ± 9.1	28.8 ± 4.6
Normal	1.0	1.7 ± 0.1	1.0	2.6 ± 0.6	14.3 ± 1.0

*Fluoride is expressed as molar percent relative to calcium.
†Mean ± SE.

Therapeutic measures included the elimination of fluoride from the drinking water, normalization of plasma calcium
and phosphate concentrations, and the use of vitamin D
analogs. Symptoms were lessened with these measures, but
several patients continued to have fractures. Four patients
have been free of symptoms or fractures since entering the
dialysis program using fluoride-free dialysate and continuing
efforts to maintain normal calcium and phosphate levels.

Case of Severest Disease. A 69-year-old man experienced
excessive frequent urge to urinate associated with pyuria in
1958. Signs of infection cleared after sulfonamide therapy,
but urinary frequency, nocturia, and polyuria persisted. The
urine was of fixed specific gravity and showed a trace of
protein. Mild azotemia appeared in 1960, followed by bone
pain, arthralgia of the knees and feet, and spontaneous
"march fractures" of both feet--a total of 13 by 1963.

Examination of the urine revealed no infection, but a
24-hour specimen showed an increased content of glucose and
amino acid nitrogen. The urine was alkaline, while the blood
showed some evidence of systemic acidosis. Blood sugar
levels were normal, azotemia was mild, and alkaline

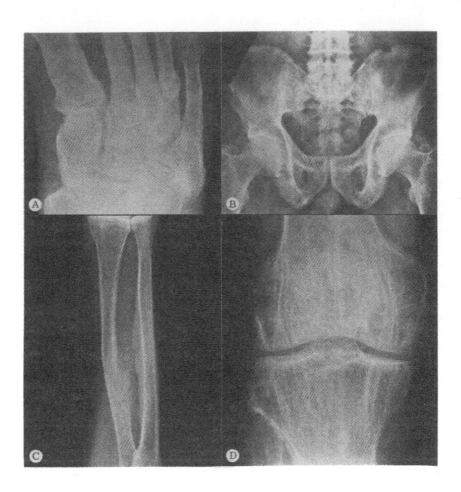

Fig. 1. A, Multiple healing fractures of metatarsals.
Note increased bone density and coarse trabecular pattern.
B, Increased density of pelvis, with coarse trabecular
pattern and calcification of sacrotuberous ligaments. C,
Calcification of interosseus ligament between radius and
ulna. D, Coarse trabecular pattern and subperiosteal new
bone formation of distal femur and proximal tibia.

phosphatase activity was elevated, but values for serum calcium, phosphate, and total protein were normal. An excretory urogram showed small kidneys. Skeletal roentgenograms showed healing fractures of the metatarsals and phalangeal bones of both feet, areas of increased bone density with a coarse trabecular pattern involving predominately the axial skeleton and calcification of interosseous ligaments, and new subperiosteal bone formation (Fig. 1).

Bone biopsy of the iliac crest showed an increase of uncalcified osteoid tissue on bone surfaces, decreased mineral density around osteocytes, low mineralization of cement lines, and much interstitium with an irregular pattern.

After treatment with oral calcium supplements and vitamin D, bone pain decreased but the patient experienced additional fractures. Osteosclerosis increased, but serum alkaline phosphatase values decreased to normal (Fig. 2). A bone biopsy specimen taken in 1968 showed healing of osteomalacia. Chemical values showed a high concentration of fluoride in serum (14 μM) and bone (4.7 to 6.5 moles of fluoride per 100 moles of calcium) and in drinking water (2 ppm or 106 μM) relative to the concentration of fluoride in the urine (78 μM).

At this point, the patient was advised to stop drinking tap water and to use only fluoride-free spring water or distilled water for both drinking and cooking. Serum fluoride concentrations decreased (to 8 μM), and for a period of approximately 8 years, the patient was relatively free of bone pain and did not experience further fractures.

In 1971, renal function temporarily deteriorated further. After peritoneal dialysis, renal function spontaneously improved. In 1974, the patient fell, sustaining a hip fracture that required internal fixation. Osteomalacia has persisted despite vitamin D therapy and reasonable control of systemic acidosis and secondary hyperparathyroidism. These findings were interpreted as representing adult Fanconi's syndrome with osteomalacia and superimposed fluorosis.

Effects of Fluoride in the Dialysate

Claims of Adverse Effects

After the introduction of home dialysis in 1964, several centers, including our own, noted a higher incidence of bone disease when patients used untreated or softened water to prepare the dialysate. One identifiable factor was the

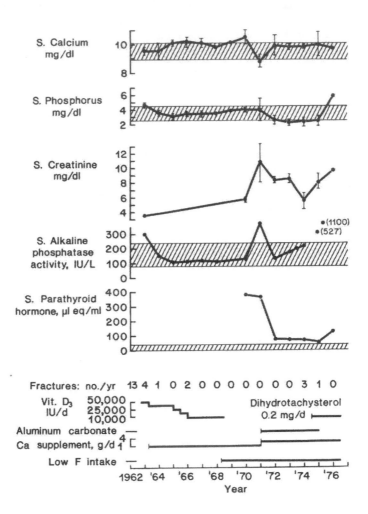

Fig. 2.   Serum chemical values, clinical response, and treatment regimen in patient exposed to fluoride before dialysis.   Hatched area = normal range.

presence of fluoride in water used at home but its absence in
water used in the centers. Unlike the bone disease usually
encountered in dialysis patients, these patients had little
evidence of hyperparathyroidism, but, instead, had histologic
evidence of severe osteomalacia. Five of the six patients
exposed to fluoridated dialysate for an average of 23 months
suffered bone pain and fractures, and three of these patients
had incapacitating symptoms. Bone biopsy specimens from five
patients exposed to fluoridated dialysate for more than 1
year were compared with those from six patients of approximately the same age, duration of azotemia, and duration of
dialysis who were dialyzed using fluoride-free dialysate.
The blood concentrations and ratios of bone fluoride to
calcium were significantly higher in patients exposed to
fluoridated dialysate (Table 5). Although the severity of
osteitis fibrosis was similar in the two groups, as reflected
by the percentage of bone surface undergoing osteoclastic
resorption, osteomalacia was significantly more severe in the
fluoridated group.

Table 5. Chemical and histologic features of
patients exposed to low and high fluoride dialysate

	Dialysate F	
	<5µM (6 patients)	>50µM (5 patients)
Hemodialysis (mo)	22.0	20.0
Fluoride		
Dialysate (µM)	1.4	52.0
Serum (µM)*	2.8	13.2
Bone F/Ca*	1.0	3.2†
Osteoid surface (%)	3.0	27.5‡
Osteoid width (µM)	17.3	37.9†
Bone formation (%)	7.0	0.3
Bone resorption (%)	29.8	26.8

*Normal: In persons drinking water with a fluoride
content of 1 ppm (53 µM), the serum fluoride is 0.7 ±
0.4 µM and the bone fluoride-to-calcium ratio,
expressed as molar %, is 1.0, which is approximately
1,000 to 2,000 ppm on an ash weight basis.
Significant difference from <5 µM group: †P<0.05;
‡P<0.01.

Case of Severest Disease. A 46-year-old man with poly-
cystic renal disease was maintained by long-term hemodialysis
(Johnson and Taves, 1974). At first, a commercial water
softener provided water containing 1 ppm of fluoride. The
serum concentrations of calcium and magnesium were maintained
in the normal or elevated range (tending to suppress parathyroid activity) and the serum phosphate values were never
below normal and, therefore, were unlikely to induce

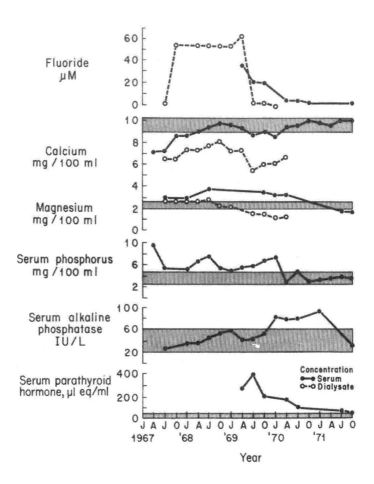

Fig. 3.   Serum and dialysate chemical values from patient exposed to fluoridated dialysate and subsequently treated by renal transplantation.   (From Johnson WJ, Taves DR.   [1974]   Exposure to excessive fluoride during hemodialysis.   Kidney Int.   5;451-454.   By permission of the International Society of Nephrology.)

phosphate-depletion osteomalacia (Fig. 3). Parathyroid hormone values also were below the average of patients on such a regimen.

Despite these measures, the alkaline phosphatase values increased steadily. Although initial skeletal surveys did not reveal abnormalities, within a year, the patient complained of chest pain and severe pain of the feet on weight-bearing. Examination of the skeleton showed general-ized demineralization and fractures of three ribs, but no evidence of subperiosteal bone resorption. In spite of a good appetite and a seemingly adequate caloric and protein intake, the patient's weight decreased 11 kg.

Because we suspected that fluoride was implicated, a deionizer was recommended. Repeated determinations of dialysate fluoride, however, revealed variable values, some levels actually exceeding the concentration in tap water. Inspection of the apparatus revealed that, because the conductivity meter was defective, the patient had been using the water meter to determine when the deionizer required regeneration. Although this practice did not result in dangerous elevation of calcium and magnesium levels in the dialysate, fluoride was eluted from the column when the deionizer was exhausted. After correction of this problem, dialysate and serum fluoride concentrations decreased rapidly.

Bone biopsy specimens obtained before and after the reduction of serum fluoride levels showed a slight decrease in the ratio of bone fluoride to calcium from 3.4 to 3.0 and some improvement in the osteomalacia and bone resorption. Despite these encouraging findings, the patient did not improve clinically and at this point was bedridden because of bone pain and fractures. After nephrectomy and splenectomy in preparation for renal transplantation, he experienced a generalized seizure and suffered additional fractures of a rib, a lumbar vertebral body, and the right femoral neck.

After successful renal transplantation, serial skeletal roentgenograms and bone densitometry measurements showed remineralization and healing of fractures. During the next 6 years, the patient was free of skeletal complications. Values for serum calcium, phosphate, and alkaline phosphatase and parathyroid radioimmunoassay concentrations returned to normal, but serum fluoride concentrations have remained elevated.

Four of six patients who were exposed to fluoridated dialysate at our institution have undergone successful renal transplantation and have been free of skeletal complaints for

as long as 7 years, although three of the four had suffered
fractures before and one immediately after transplantation.

Discussion

A number of studies have supported the conclusion that
exposure to fluoridated dialysate increases the risk of
symptomatic osteomalacia. Posen and co-workers (1971) noted
an increased incidence of bone disease in the city of Ottawa,
where fluoridated water was used to prepare the dialysate;
and Cordy et al. (1974) reported a similar experience when
patients who underwent dialysis at home used fluoridated
water. The incidences of bone pain and fractures in Montreal
(Cordy et al., 1974) were lower than those in our experience
(Jowsey et al., 1972) and those reported in Ottawa (Posen,
Marier, and Jaworski, 1971). Biopsy specimens from the iliac
crest of four patients in Montreal who had the highest ratios
of fluoride to calcium in bone were studied by Lough et al.
(1975), using phase contract and electron microscopy. These
biopsy specimens were compared with specimens from patients
of similar age, sex, and duration of treatment who had been
using fluoride-free dialysate. The studies differed in not
having any adolescents and in dialysis mainly being done in a
center rather than at home. In the nonfluoridated group, the
findings were those of hyperparathyroid bone disease. On
light microscopy, an increase was noted in the proportion of
bone covered by osteoid in the fluoridated group. Cortical,
cancellous, and periosteal bone showed abnormal mosaic
patterns and staining characteristics. The number of osteoblasts was increased, and the canaliculi were in disarray.
Electron microscopy confirmed an increase in osteoblasts and
showed that their cytoplasmic processes were tangled and
disarrayed. Collagen fibers were of normal diameter and
periodicity but were loosely and irregularly aligned in
bundles compared to the nonfluoridated group. Similar
findings have been reported in cattle and dogs exposed to
high fluoride intake--for example, osteoblastic cytoplasmic
processes that produce collagen develop a haphazard,
rootlike appearance, and the collagen that is produced is
excessive in amount and lacks orderly orientation (Johnson,
1965).

Consistent with the thesis that fluoride contributes to
osteomalacia is the report of Posen and co-workers (1972),
who compared the incidence of osteodystrophy in patients
dialyzed either with deionized water or with fluoridated tap
water. Deionization was associated not only with less
symptomatic osteomalacia but also with healing of prior
osteomalacia. Because other elements as well as fluoride
were removed by deionization, the authors suggested that

fluoride may contribute to the development of osteomalacia, but they could not exclude other factors.

In a study of the ionic composition of bone from uremic patients, Parsons and co-workers (1971) demonstrated uneven distribution of fluoride throughout the skeleton and concluded that the higher concentration of fluoride in certain areas was responsible for the increased osteoid in those locations. Siddiqui et al. (1970) reported that serum fluoride concentrations increased with time on dialysis and found a significant correlation between the severity of bone disease and the concentration of serum fluoride. Bone biopsy specimens from 17 patients revealed that the highest fluoride concentrations were present in those with severe osteomalacia. Kim and co-workers (1970) demonstrated a high content of fluoride in cortical bone of patients with renal disease whether or not they were treated with hemodialysis. The ratios of fluoride to calcium, however, were highest in the patients using fluoridated dialysate.

However, other studies have cast doubt on the hypothesis that the use of fluoridated dialysate causes more osteomalacia. Siddiqui and co-workers (1971) compared bone disease in patients from two cities in England: in patients from Birmingham, symptomatic osteodystrophy is uncommon and usually associated with osteitis fibrosa, while in patients from Newcastle, osteodystrophy is common and associated with multiple fractures but without roentgenographic evidence of osteitis fibrosa. Incrimination of fluoride in tap water as a cause of severe bone disease was not possible because both centers used fluoridated dialysate. Bone biopsy specimens were not examined in this study. Speculation as to the presence of some other toxic solute, as yet unidentified, was the best available explanation for the difference in attack rates of bone disease between the two cities. Oreopoulos and co-workers (1974) did a prospective double-blind study using deionized water to prepare the dialysate. In one group of patients, sodium chloride was added, while in the other, sodium fluoride was added to provide fluoride concentrations of 1 ppm in the final dialysate. Bone biopsy specimens obtained before treatment and after a year failed to show significant differences in the two groups except for increased sclerosis in the fluoride group. They concluded that dialysate fluoride is not the primary cause of progressive osteomalacia, at least during the first 2 years of dialysis.

Conflicting reports regarding the effects of fluoride in patients dialyzed using untreated tap water containing a high concentration of fluoride for preparation of the dialysate may be explained in a number of ways. Distinctive

roentgenographic changes associated with fluorosis may not be
seen even after prolonged exposure, perhaps because renal
failure alters the capacity of the dialyzed patient to
respond by increasing bone production. Clinical manifestations of bone disease such as arthralgia, bone pain, and
pathologic fractures depend on many variables other than
fluoride. Therefore, the presence or absence of symptoms,
fractures, or roentgenographic changes also is not a reliable
index of a fluoride effect. In most instances, the only
indications of the positive role of fluoride are an elevation
of fluoride in blood and bone and the histologic changes
demonstrated on bone biopsy. The significance of an elevation
of serum fluoride alone may be ambiguous because
increased bone resorption as a result of uncontrolled
hyperparathyroidism also may lead to an elevation of serum
fluoride. However, failure of the double-blind study to show
a statistically significant increase in osteomalacia does not
rule out some effect of fluoride. A small but not statistically significant difference in the expected direction was
noted after a longer time (Oreopoulos, 1977). The control
patients in the Oreopoulos study did not provide a clear
contrast with the experimental group. Two of nine patients
in the low fluoride group had long predialysis and prestudy
exposure to fluoride. As a consequence, serum and bone
fluoride concentrations of the two groups overlapped considerably. Also, the serum and bone fluoride concentrations
they reported are lower than those noted in the Mayo Clinic
(Jowsey et al., 1972) and the Montreal (Cordy et al., 1974)
studies. The limitations of Oreopoulos' study may explain the
failure to obtain a clear difference between the groups
studied.

Concluding Remarks

The available evidence suggests that some patients with
long-term renal failure are being affected by drinking water
with as little as 2 ppm fluoride.   All of the patients
showed increased bone density, and two showed calcification
of interosseous ligaments which is thought to be diagnostic
of skeletal fluorosis (Stevenson and Wilson, 1957). The
average concentration of fluoride in bone of 4.4 moles of
fluoride per 100 moles of calcium is equivalent to 9,000 ppm
of fluoride on an ash weight basis and is in the middle range
of the values that have been reported for advanced fluorosis
(Hodge and Smith, 1965; Jolly, Singh, and Mathur, 1969). The
excessive osteoid formation seen in these patients is probably accentuated by fluoride.

With regard to the patients exposed to fluoridated
dialysate, the presumption should probably be that fluoride

has a definite but minor role in the development of osteomalacia.

The meaning of these findings for community fluoridation will depend on whether or not further work will clearly
show adverse effects in patients with renal failure drinking
water with a concentration of 1 ppm of fluoride and whether
these effects can be easily avoided. The finding of adverse
effects in patients drinking water with 2 ppm of fluoride
suggests that a few similar cases may be found in patients
imbibing 1 ppm, especially if large volumes are consumed, or
in heavy tea drinkers and if fluoride is indeed a cause.

It would seem prudent, therefore, to monitor the
fluoride intake of patients with renal failure living in
high fluoride areas. The serum concentration may indicate
whether the patient should be advised to drink low fluoride
water and will provide a check regarding compliance.
Tentatively, a shift to low fluoride water should be made
before the serum fluoride concentration reaches 5 µM, since
evidence of fluorosis has been reported when the average
serum concentrations of fluoride are 8 µM (Leone et al.,
1955; Singla, Garg, and Jolly, 1976).

In patients maintained by dialysis for long periods,
fluoride-free dialysate should be used. There is insufficient evidence at this point to recommend the use of
fluoride-free drinking water for all patients with renal
disease.

References

Berman, L.B. and Taves, D.R. (1973) Fluoride excretion in
    normal and uremic humans (abstract). Clin.Res., 21;100.

Carlson, C.H., Armstrong, W.D. and Singer, L. (1960) Dis-
    tribution and excretion of radiofluoride in the human.
    Proc.Soc.Exp.Biol.Med., 104;235-239.

Cordy, P.E. et al. (1974) Bone disease in hemodialysis
    patients with particular reference to the effect of fluo-
    ride. Can.Med.Assoc.J., 110;1349-1353.

Hein, J.W. et al. (1956) Distribution in the soft tissue of
    the rat of radioactive fluoride administered as sodium
    fluoride. Nature, 178;1295-1296.

Hodge, H.C. and Smith, F.A. (1965) In Fluorine Chemistry,
    Vol. 4, edited by J.H. Simons. New York, Academic Press,
    pp. 152, 155, 171, 443.

Hodge, H.C. and Smith, F.A. (1968) Fluorides and Man.
    Annu. Rev. Pharmacol.,8;395-408.

Hosking, D.J. and Chamberlain, M.J. (1972)   Studies in man
    with ^{18}F.   Clin. Sci.,42;153-161.

Johnson, L.C. (1965)   Histogenesis and mechanisms in the
    development of osteofluorosis.   In Fluorine Chemistry.
    Vol. 4.   Edited by J.H. Simons. New York, Academic Press,
    pp. 424-441.

Johnson, W.J. and Taves, D.R. (1974)   Exposure to excessive
    fluoride during hemodialysis.   Kidney Int.,5;451-454.

Jolly, S.S., Singh, B.M., and Mathur, O.C. (1969)   Endemic
    fluorosis in Punjab (India).   Am. J. Med., 47:553-563.

Jowsey, J.O.et al.(1972)   Effects of dialysate calcium and
    fluoride on bone disease during regular hemodialysis.
    J. Lab. Clin. Med., 79;204-214.

Jowsey, J.O., Riggs, B.L. and Kelly, P.J. (1978)   Fluoride
    in the treatment of osteoporosis.   (This monograph).

Juncos, L.I. and Donadio, J.V., Jr. (1972)   Renal failure
    and fluorosis.   J.A.M.A., 222;783-785.

Kaye, M. et al. (1960)   Bone disease in chronic renal fail-
    ure with particular reference to osteosclerosis.
    Medicine (Baltimore), 39;157-190.

Kim, D. et al. (1970)   Bone fluoride in patients with
    uremia maintained by chronic hemodialysis.   Trans. Am.
    Soc. Artif. Intern. Organs, 16;474-478.

Leone, N.C. et al. (1955)   A roentgenologic study of a
    human population exposed to high-fluoride domestic water:
    a ten-year study.   Am. J. Roentgenol., 74;874-885.

Lough, J. et al. (1975)   Effects of fluoride on bone in
    chronic renal failure.   Arch. Pathol., 99;484-487.

Nielsen, E. et al. (1973)   Fluoride metabolism in uremia.
    Trans. Am. Soc. Artif. Intern. Organs, 19;450-455.

Oreopoulos, D.G. (1977)   Personal communication.

Oreopoulos, D.G. et al. (1974)   Fluoride and dialysis osteo-
    dystrophy: results of a double-blind study.   Trans. Am.
    Soc. Artif. Intern. Organs, 20;203-208.

Parsons, V. et al. (1975)   Renal excretion of fluoride in
    renal failure and after renal transplantation.   Br. Med.
    J., 1;128-130.

Parsons, V. et al. (1971)   The ionic composition of bone
    from patients with chronic renal failure and on RDT,
    with special reference to fluoride and aluminum.   Proc.
    Eur. Dial. Transplant Assoc., 8;139-147.

Posen, G.A. et al. (1972) Comparison of renal osteodystrophy in patients dialyzed with deionized and non-deionized water. Trans. Am. Soc. Artif. Intern. Organs, 18;405-409.

Posen, G.A., Marier, J.R. and Jaworski, Z.F. (1971) Renal osteodystrophy in patients on long-term hemodialysis with fluoridated water. Fluoride, 4;114-128.

Prosser, D.I. et al. (1970) The movement of fluoride across the cuprophane membrane of the Kiil dialyser. Proc. Eur. Dial. Transplant Assoc., 7;103-109.

Rao, T.K.S. and Friedman, E.A. (1975) Fluoride and bone disease in uremia. Kidney Int., 7;125-129.

Sauerbrunn, B.J.L., Ryan, C.M. and Shaw, J.F. (1965) Chronic fluoride intoxication with fluorotic radiculomyelopathy. Ann. Intern. Med., 63;1074-1078.

Siddiqui, J.Y. et al. (1970) Serum fluoride in chronic renal failure. Proc. Eur. Dial. Transplant Assoc., 7;110-117.

Siddiqui, J.Y. et al. (1971) Fluoride and bone disease in patients on regular haemodialysis. Proc. Eur. Dial. Transplant Assoc., 8;149-159.

Singla, V.P., Garg, G.L. and Jolly, S.S. (1976) Non-skeletal phase of chronic fluorosis--the kidneys. Fluoride, 9;33-35.

Smith, F.A., Gardner, D.E. and Hodge, H.C. (1955) Investigations on the metabolism of fluoride. III. Effect of acute renal tubular injury on urinary excretion of fluoride by the rabbit. Arch. Industrial Health, 11;2-10.

Stevenson, C.A. and Wilson, A.R. (1957) Fluoride osteosclerosis. Am. J. Roentgenol., 78;13-18.

Taves, D.R. and Guy, W.S. (1978) Distribution of fluoride among body compartments. (This monograph.)

Taves, D.R. et al. (1968) Hemodialysis with fluoridated dialysate. Trans. Am. Soc. Artif. Intern. Organs, 14;412-414.

Wallace-Durbin, P. (1954) The metabolism of fluorine in the rat using F^{18} as a tracer. J. Dent. Res., 33;789-800.

Wootton, R. (1974) The single-passage extraction of ^{18}F in rabbit bone. Clin. Sci. Mol. Med., 47;73-77.

Young, R.A., Van der Lugt, W. and Elliott, J.C. (1969) Mechanism for fluorine inhibition of diffusion in hydroxyapatite. Nature, 223;729-730.

This study was supported in part by Research Grant AM-8658 from the National Institutes of Health, Public Health Service.

13

Claims of Harm from Fluoridation

Donald R. Taves

There are some well-publicized claims of harm from fluoridation, but these have not been adequately confirmed from a scientific point of view. This is not to say, however, that they have been adequately refuted either. Scientists are generally skeptical of important claims unless the data: (1) come from well-controlled experiments designed to rule out bias and selection factors as causes; (2) are confirmed independently by others; and (3) make sense in the light of other knowledge. On the other side of the coin, scientists are generally skeptical about claims of no effect, especially when confidence intervals for the reported findings are wide or have not been established. How narrow the confidence intervals need to be is a subjective decision which can change with time and depends on cost-benefit and risk-benefit assessments. It is, therefore, appropriate to review the claims of harm, to examine why they have not been generally convincing to scientists, and to note what might feasibly be done to reduce the chance that future investigators will say we just did not look carefully enough.

Sensitivity Reactions

Waldbott has ascribed a large number of symptoms to the ingestion of 1 ppm F in water or of 1 mg F as a test dose, and his most convincing evidence of this claim was published in 1962. Two of these symptoms - gastrointestinal distress and joint pains - have been reported in a few patients in studies when a daily dose of 20 or more mg F was given for treatment of bone disease (Shambaugh and Sundar, 1969; Rich et al., 1964). Polyuria has become accepted as an effect of fluoride when the serum fluoride concentration is above 50µM, a level which can be produced by the metabolism of methoxyflurane (Van Dyke, this monograph). The remaining symptoms (listed in order of frequency): stomatitis, headache, backache, weakness or lethargy, parathesias, tenderness of muscles, swelling or edema, dysuria, limitation of motion, loss of memory, muscle fibrillation, conjunctivitis, rhinitis, tinnitus, vertigo, bleeding gums, urticaria and blurred vision, have not been accepted by most physicians as being caused by chronic intake of fluoride at any dose level. The lack of acceptance persists in spite of Waldbott's report that most of these symptoms were reproduced not only by inadvertent patient exposure or physician challenge but by double-blind challenge with 1 ppm F. Since 1962 the main addition has been the claim that Chizzola maculae (dime-sized bruise-like lesions of the skin) are caused by fluoride exposure from industrial pollution (Steinegger, 1969; Colombini et al., 1969) and from fluoridated water (Waldbott and Steinegger, 1973).

Waldbott's claim of effects with 1 mg fluoride has been dismissed on four grounds: (1) that he is the only one to report symptoms at this low dose (Schlesinger, 1970); (2) that the solutions given to the patients for addition to their drinking water at home have been sufficiently concentrated to be distinguished by taste and that the curious patient can break the code and jeopardize a double-blind test (Royal College of Physicians, 1976); (3) that the symptoms are too varied to make sense (Schlesinger, 1970); and (4) that this type of sensitivity has not been reported among the billions of tea drinkers in the world (Ericsson, 1970). The Chizzola maculae claim has been dismissed on the basis of inconsistent temporal and spacial distribution of the lesions (Hodge and Smith, 1977). Most of these counter arguments do not inspire much confidence, and the first one is not even factual.

Waldbott is not the only one who has described patient syndromes which were explained as intolerance to fluoride. Douglas (1957) tested 32 patients from a group of 133 with histories suggestive of sensitivity to fluoride-containing dentifrices. He states that none were able to complete a series of six trials alternating between fluoride and non-fluoride toothpaste because of intolerance, mainly in the form of ulcerations of the mouth. Feltman and Kosel claimed that, among pregnant mothers and their children, 1% (at least four) reacted adversely to 1 mg fluoride in tablet form. They claim to have established (by means of placebo) that it was the fluoride, rather than the binder, which caused the adverse effects (Feltman, 1956; Feltman and Kosel, 1961). Shea, Gillespie, and Waldbott (1967) reported seven cases of patient improvement after discontinuing vitamin drops or toothpaste containing fluoride. In the cases involving toothpaste, the associated cation is not stated. Stannous fluoride is commonly used in toothpaste. Therefore, sensitivity to tin or other constituents, rather than to fluoride, can

not be ruled out. Petraborg (1974) reported seven case histories of alleged fluoride sensitivity, but the patients were
not tested, so the evidence is weak. Zamfagna (1976) reviewed the literature and added two cases, one of which was "confirmed by positive challenge test". Grimbergen (1974) clearly described a case with ulcers in the mouth which was confirmed by double-blind test. He also gave a "preliminary"
report on "60 patients, selected from a group of about 300
individuals who had suspected ill effects from fluoridated
water." Waldbott used a concentration of NaF (2.2 mg/15 ml
or 1 mg F/ml) which is probably too low to be identified by
taste. The Royal College of Physicians (1976, p. 63) review
stated that 1 mg F/15 ml has a distinctive taste. Taves (unpublished, 1976) however, found that four out of five people
could not tell the difference between fluoride-free distilled
water and distilled water with 1 mg F/15 ml. It should be
noted that care has to be exercised with the water. If the
2.2 mg NaF/15 ml solution is stored in plastic containers or
is made from a more concentrated stock solution which has
been stored in plastic, sufficient taste can be present from
the plastic to make it distinctive when compared to a solution stored in glass or made with fresh distilled water.

The variability of symptoms reported by Waldbott and the
lack of a theoretical basis for their production are not
secure arguments but do carry some weight. Chronic poisoning
can have quite varied symptoms but the frequency of signs and
symptoms increases with dose. For instance, see the symptoms
for the drug isoniazid (Goodman and Gilman, p. 197, 1975). A
theoretical physiological basis is not impossible to con-
struct. Quissell and Suttie (1972) have described cells
which are resistant to high doses of fluoride because of
their ability to pump fluoride out of the cell interior
(Repaske and Suttie, this monograph). Sensitive tissues or
subjects may be those lacking this ability.

Cook (1972) as well as Waldbott (1962) described a case
of arthritis which improved when heavy tea consumption, hence
fluoride, was eliminated. The general lack of reports among
tea drinkers is an argument made questionable by a recent
report (Finn and Cohen, 1978) which indicates that coffee and
tea may be responsible for some cases of undiagnosed symptoms.
The authors do not discuss the possibility that the cause of
the symptoms is fluoride. The active ingredient could be the
xanthines which are common to both beverages. The point of
interest is that these beverages were taken for granted to
such an extent that both patients and physicians had missed
their significance.

The evidence for Chizzola maculae from fluoride has been
extensively reviewed by Cavagna and Bobbio (1970) and by

Hodge and Smith (1977). Their conclusion that Chizzola maculae is not due to the fluoride is based on three main points. First, the reported incidence of lesions disappeared from 1938 to 1965 even though the aluminum factory continued operation. Cavagna and Bobbio state that the emissions during that period of time were greater than in October 1968, when the incidence of skin lesions in children was 98%. Waldbott and Steinegger (1973) state that the operations were discontinued during that period of time, but one of their references, Colombini et al. (1969), state that the decrease in incidence was associated with the temporary halting of operations in order to install equipment for the treatment of the waste gases and that the increase from 1965 on was due to a marked increase in production.

Second, Cavagna and Bobbio quote a commission report which studied seven surrounding towns not subject to the air pollution from the aluminum plant and found 36% to 52% of the children in these towns affected at the same time that 49% of the children were affected in Chizzola. This is in marked contrast to the statements of Waldbott and Steinegger and Columbini et al. that the incidence was nearly zero in the surrounding towns.

Third, urinary fluoride concentrations were not higher in Chizzola than in the surrounding towns (Waldbott and Steinegger, 1973; Hodge and Smith, 1977; Cavagna and Bobbio, 1970); therefore, the effect, if due to fluoride, presumably is not a parenteral effect. The absence of lesions in communities with sufficient fluoride pollution of the air to affect urine fluoride concentrations and cause dental mottling would suggest that special circumstances must be involved if fluoride emission is the culprit. The discrete round lesions (see pictures in Colombini et al., 1969), usually on the extremities of children and women, suggest that particles with absorbed chemicals might have been lodging on the exposed skin and causing local damage. Acid adsorption would convert any fluoride in the particle to HF or possibly H_2SiF_6 which would release HF upon contact with the skin. The HF molecule is highly penetrating and might cause such type damage. The special circumstances would then not be hard to imagine. Particle size and meterological conditions, particularly rainfall, might make considerable difference in the fallout. Research in animals (Olivo et al., 1968) quoted by Cavagna and Bobbio, and Colombini et al. (1969) has been done by feeding experiments only so that the absence of skin lesions is not conclusive. Skin testing with fallout particles would appear to be in order but these do not appear to have been done.

Waldbott and Steinegger (1973) claimed that Chizzola

maculae is caused by drinking fluoridated water; however,
they did not offer evidence of its production by fluoride
challenge. The possibility that ingested fluoride causes
this type of discrete skin lesion would surely have been reported among women taking large amounts of fluoride for osteoporosis. The physicians might not notice the lesions but how
could these escape the women?

Perhaps the most disturbing aspect of the "sensitivity"
reports is the failure to pay attention to contrary evidence.
The supposition in the Chizzola maculae case is that those
near the factory were being exposed to more fluoride than
more distant villagers. Therefore, when the findings show
that the urine levels are no different, the basic supposition
should be questioned. This sort of failure to question and
be disturbed by such discrepancies indicates that the investigator may tend to dismiss contrary evidence that comes from
patients as well.

There are some additional reasons for being skeptical
about the claim of effects from low doses of fluoride. If
Feltman and Kosel's estimate of 1% intolerant individuals is
correct, there should have been more reports of adverse effects in the studies in which fluoride tablets were given to
at least 10,000 school children (O'Meara, 1968). Also, as
methoxyflurane anesthesia for surgery typically causes serum
fluoride content to increase 30-50 times normal (Fry et al.,
1973), there should have been striking cases of such intolerance in an estimated 12 million patients who have received
methoxyflurane (NAS-NRC, 1971). There have not been reports
of intolerance from people who move into and out of the
numerous towns with naturally high fluoride levels in the
water supplies. Opportunities for such discovery existed
before any bias for or against fluoridation. A patient,
having gone from doctor to doctor and probably being labeled
a "crock", who finds a sympathetic doctor who "knows" that
fluoride is the source of his problems, is going to try very
hard to support this explanation. Expectation plays a very
large role in what subjects experience when given placebos
(such as capsules of sugar). Green reported a wide variety
of symptoms (gastrointestinal, heartburn, drowsiness, blurred
vision, dizziness, dry mouth, palpitation, urinary frequency
and vomiting) among half of 50 professional people given
placebos. The professionals, expecting that they might be
getting an active drug, naturally worried about the side effects. Patients tended, on the other hand, to have a decrease in symptoms when given placebos. (Green, 1964)

Most of the above counter arguments are based on passive
observations; so while it seems unlikely to most scientists
that fluoride is causing adverse effects at 1 ppm F, active

study is desirable.

The active study which should be done is to systematically test patients with unexplained illness in those clinics which are likely to see patients with the common symptoms described by Waldbott. These patients should be told only that their problem might be something in the water so that bias for or against fluoridation would not complicate the testing. Those who improved on low-F bottled water should then be tested with double-blind addition of NaF or NaCl concentrate added to the low fluoride water. The concentrate should be no more than 2.2 mg NaF/15 ml. Patients who responded positively to this test ideally would then be tested in a clinical research unit where the double-blind tests could be supervised more closely and the fluoride distribution and renal clearance rates studied. Failure to find distinctive objective features would not rule out the validity of the subjective response, but demonstration of distinctive objective responses would be much more effective in convincing other scientists that a real effect was being observed. This might also lead to an understanding as to the how or why, which would really be effective in making a case. To make the studies even more convincing the investigator should not have taken either of the political positions in regard to fluoridation.

Mongolism

The possibility that mongolism is caused by fluoride in the drinking water stems from reports by Rapaport (1956, 1959, 1963) in which he observed a dose-related association between the number of cases of mongolism and the concentration of fluoride in the water. From towns with less than 0.1 ppm to those with 1.0 to 2.6 ppm, the increase was nearly three-fold.

Rapaport's original report was criticized on the basis that the cases were not assigned by the place of residence of the mother but by the place of delivery (Russell in Hodge and Smith, 1965). This assignment allowed women from farms with shallow wells and low fluoride water to be included with those from towns with deep wells and high fluoride water. Unless the denominators of his rates did not include the rural areas, the effect of the mistake would be a conservative error on Rapaport's part and Russell's argument would have no weight. If Rapaport made the more serious error of non-correspondence of his numerator (cases of Mongolism) to his denominators (populations), Russell missed the important weakness of Rapaport's work.

Both men did subsequent work in which place-of-usual residence was considered but their results are conflicting and not decisive. Russell is also quoted by Hodge and Smith (1965) and describes redoing part of the work to correct the place-of-residence fault by considering a limited number of more homogenous counties and finding that the rates showed too little difference to be statistically significant. This is not convincing because reduction of the sample size automatically decreases the statistical significance of Rapaport's data, regardless of the validity of his results. Russell's failure to publish his results in the usual way also leaves something to be desired.

Rapaport's next study (1959) acknowledged consultation with Russell in regard to limiting the study to the State of Illinois and cities of 10,000 to 100,000 and he indicated that he used the place of habitual residence of the mother. This time he used the number of live births for denominators rather than population which is much better; however, he did not state how he obtained the denominators for his rates. Rapaport implied, however, that he obtained numerator and denominatory by city. In which case, it would not have made sense for Russell to have used county figures, which refer to populations not as homogenous for fluoride intake as do city figures. In any event, Rapaport clearly did not consistently use ideal rates. He divided both the number of Mongoloid births in mothers over and under age 40 by the number of births of all ages; thus, these are not age-specific rates which would be best for comparisons of high and low fluoride exposures. This failure makes his conclusion, that the younger mothers were more affected by fluoride, questionable, although it is hard to see how the lack of age-specific rates could artificially produce both high total Mongoloid rates and a low average maternal age with Mongoloids in the fluoride areas.

Needleman et al. (1974) studied the rates of Mongolism in Massachusetts residents between 1950 and 1967 by classifying the mother as having used or not used fluoridated water. This was based on the status of the water supply of the residence nine months prior to delivery as given (presumably) at delivery. There are no naturally fluoridated communities in Massachusetts. Therefore, all fluoride exposures would have been for relatively short (few years) time periods. Needleman et al. concluded that more than a 25% increase in Mongolism from short-term exposure could be ruled out. They noted that since the first meiotic division of the ovum starts in utero, a long period of induction is not impossible.

Three intensive case-finding studies in Britain (Berry, 1958, and two unpublished ones cited by the Royal College of

Physicians, 1976) with different fluoride concentrations in the water have not shown such an association. Heavy tea drinking in England (Cook, 1970) might have obscured differences in the British studies. However, the absolute rates were similar to those in the intensive case-finding study in Massachusetts. Therefore, for Rapaport's hypothesis to be maintained, an explanation as to why the British rate does not reflect increased consumption of fluoride from tea drinking would be necessary.

Erickson et al. (1976) reported on two separate studies from which they concluded that there is no evidence for an association between fluoride and Mongolism. However, their conclusion is weakened because of low ascertainment in one of the studies similar to that which the Royal College of Physicians considered to have made Rapaport's studies unreliable. The other study showed evidence of good ascertainment but the overall (crude) rate was 18% higher in the fluoride-exposed group. This is not statistically significant by itself but, of course, can not be used to rule out a potential effect of that amount. Low ascertainment of the total number of cases occurs because the condition frequently is not recognized at birth and consequently is not recorded on the birth certificate. In order for low ascertainment to cause errors there must be some difference in how carefully these cases were reported from the two areas. Burgstahler (1977) claimed that Erickson's findings confirmed those of Rapaport, namely that younger mothers are most affected, but this argument is based on selective use of the data. The only age subgroup to show a statistically significant difference (considered independently) was the 35- to 40-year mothers and the effect was in the opposite direction. If young mothers are considered to be those under 40, as Rapaport (1963) did and as Burgstahler (1975) did, the difference in rates is not statistically significant for either of Erickson's studies, ψ^2 less than 1. If young mothers are considered to be those under 35, as Burgstahler did in 1977 (personal communication), there is a statistically significant difference ($\psi^2 = 6.8$) in one study (high ascertainment, higher Mongolism rates in the fluoride-exposed). The older mothers in the low ascertainment study have a lower rate, $\psi^2 = 5.06$. The other age groups have the same trends but the ψ^2's are less than 1.5.

In conclusion, the case for the claim that fluoridation leads to increased Mongolism is based on questionable and selected data and the case against the claim is based on studies with short-term exposure, and on data which can not rule out an increase for long-term exposure, particularly in young mothers. Further study would, therefore, seem to be in order on large populations with life-time exposure.

Mutagenesis

There are a number of papers suggesting or claiming that fluoride is mutagenic, but relatively little attention has been paid to them. This lack of attention probably stems from the fact that the published evidence has been in plants and Drosophila, primarily with high doses of hydrogen fluoride (HF) rather than F ion. As noted below, the possibility of mutagenesis due to HF is potentially important in cancer of the stomach. Ingested fluoride ion can become HF in the stomach because the pK of HF is 3.18 and pH of the stomach without food is generally about 1. Although stomach cancer rates show no indication of relationship to fluoridation in the United States, the much higher stomach cancer rates in Japan are related to intake patterns which are compatible with the hypothesis that fluoride is the crucial factor involved (Taves, 1977). Therefore, the work in plants and Drosophila with NaF and HF will be reviewed after considering the work of Mohamed and Chandler (1977) who reported that 1 ppm NaF (0.45 ppm F) in the drinking water caused damage to the chromosomes of bone marrow cells and spermatocytes in mice.

Mohamed and Chandler's experiment consisted of feeding mice a low fluoride diet and drinking water (distilled) containing varying concentrations of sodium fluoride for two time periods, 3 and 6 weeks. The concentrations were 0, 1, 5, 10, 50, 100, and 200 ppm NaF. Sixty-four mice, plus eight baseline controls, were put on low-fluoride diet and given distilled water to drink for one week prior to the start of the experiment. The resulting data consisted of food intake, water, consumption, fluoride concentration of the bone ash, and cytological information on chromosomes from bone marrow cells and spermatocytes.

The draft of the paper available for review by the National Academy of Science in 1977 contained errors and omissions which were disturbing. The bone values were less than any previously published for low fluoride diets even though some of Mohamed and Chandler's mice received 200 NaF in their water for six weeks! Also, the original Table V showed many percentages which could not be derived from any whole number of abnormalities given the number of cells they said they examined. The published paper (Mohamed and Chandler, 1977) contains over 70 changes in the Table V. How the decisions were made regarding the number of slides prepared per group and how many cells were counted on each slide or whether this was done blindly, was not indicated.

The largest change in frequency of chromosome abnormality

occurred between the 0 and 1 ppm NaF groups with little or no difference in the bone fluoride concentrations. This is so unusual a dose response that extraneous factors need to be checked. One factor which needs investigation is the amount of metal ions picked up from the metal drinking tubes with and without fluoride in the distilled water. A difference in metal intake could be important in a diet which may be marginally iron-deficient. This diet was the same as that used by Taylor et al. (1961) which Tao and Suttie (1976) showed can be marginally iron-deficient. If one ignores the distilled water group, and considered the 1 ppm NaF animals as controls there would be little change in chromosomal fragments. There would still be a doubling of "ball metaphase", rings, and bridges but there may be some difficulty in counting these.

Mohamed's earlier work was done on plants and also has some methodologic weakness. The work on onion root tips (Mohamed et al., 1966a) and tomatoes (Mohamed et al., 1966b) was done without the use of random number to code the slides and although his study on corn (Mohamed, 1970) utilized random assignment of the slide numbers prior to reading, this is not mentioned in the paper (Personal communication, A. Mohamed). An unpublished study by P. Temple and L.H. Weinstein, personal communication, 1976) confirmed the increased frequency of bridge sand fragments of chromosomes in onion root tips when grown in 10^{-2} M fluoride, but did not confirm the observation of ball metaphase.

While Mohamed (1977) did not cite confirming studies of his early work in plants, there are at least four which are purported to confirm it. Bale and Hart (1973a) studied the effects of 10^{-2}, 10^{-4} and 10^{-6} M NaF and HF on the 12 to 72-hour growth of barley seedlings. They found that the percentage of chromosomal abberations in the roots approximately doubled with the 10^{-6} NaF solution and doubled again at 10^{-2} M. The concentration of 10^{-6} M is only 0.02 ppm, and it would take special precautions to avoid introducing that amount from glassware contamination alone (Husdan et al., 1976). This extremely low concentration is to be contrasted to the concentrations and units used by Bale and Hart in a companion article (1973b). The later study involved exposure (1 hour) of coleoptiles (rather than the whole seedling) to 1, 4 and 6% sodium fluoride or 0.25, 1 and 1.5 M; i.e., the lowest solution concentration is 25X more concentrated than the most concentrated solution of the previous study. The 57th edition of the Handbook of Chemistry (1976) lists the solubility of NaF as 4.22 gm per 100 ml, a 4.2% solution, and that is the highest concentration that we have been able to make in our laboratory (Taves, unpublished). Therefore, the 6% figure must involve an error. They noted increased chromosomal breaks (did not state whether or not the reading of the slides was blind) but they did not find any evidence of chlorophyll mutation. The marked effects found that concentrations so low as to require special handling and the shift to unreasonably high fluoride levels and to different concentration units make it necessary to have these studies verified before accepting the results as confirming Mohamed's work in plants. As of September 1977 these articles had been cited only by two review articles so apparently there has been no duplication as yet. Bale and Hart (1973a) refer to two theses from Texas A. & M. University as confirmation of Mohamed's studies on mutagenesis (Abid, 1967; Mouftah, 1968). No journal publications have been found on this work by these authors.

The only published second-generation study to provide evidence of inheritable mutation was done by Mohamed (1968). Abnormal numbers and shapes of the cotyledons occurred in the treated groups about three times more often than the 4.7% frequency in the controls. Fasiated petioles, wiry and plumuless plants were noted only in the treated groups but the frequency was irregularly related to the duration of exposure. Dwarf plants and double stalks increased from 0.8% in the controls to 5% in the treated plants with uniform increases matching length of exposure in four of the five time periods. With no criteria listed, the descriptions of damage are not based on specific notations and the apparent lack of blind reading leaves the data questionable (Miller, M.W.,1977, personal communication). Mohamed's article stated that work is in progress to determine whether the effect is due to minute chromosomal changes or to changes within the gene. However, Bale and Hart's reports are the only experimental papers to cite this article through the September 1977 issue of Citation Index, suggesting that direct evidence of transmission of a mutation to progeny may not have been confirmed. Temple and Weinstein (1976) were unable to confirm the findings with the same or a different strain of tomato. The negative results with the same strain are of limited importance, however, because of a very low rate of seed production under the conditions used.

There have been studies on Drosophila by several different laboratories with a variety of conclusions. Mohamed (1971) claimed to have proven that HF in a concentration too low to cause death is mutagenic in Drosophila melanogaster. His claim is based on genetic analysis of the progeny of flies exposed for 6 to 12 hours to air which had been bubbled through 2.5% HF. The analysis involved observation of the ratio of the homozygous to heterozygous (second chromosomes) offspring from F_1 generation sib matings. The tester females were heterozygous for a recessive lethal gene. Therefore,

F_2 generations could only be homozygous for the paternal chromosomes in question; i.e., 33% would be expected. These F_1 sibs were selected on the basis of marker genes to have the same paternal and maternal genes. The control groups showed 34.68 ± 0.86% homozygous and the treated groups showed 26.84 ± 0.09 to 25.52 ± 1.54% homozygous, a clearly significant difference. The author claimed a statistically significant difference between the treated groups, but this is true only if the data in the table are in error and standard deviations are reported rather than (as stated) standard errors. The small dose effect may be due to discontinuance of the administration of HF after one hour on the erroneous assumption that the concentration would stay constant for the remaining time of the exposure period. Actual measurement of the concentration of fluoride in the air of the chamber would be helpful. Also, a tighter case could be made by mating the males prior to treatment to show that their genes are in fact normal before exposure to HF, rather than to rely on sibling controls.

Mukherjee and Sobels (1968) reported that the injection of 1 mM NaF (unstated volume) in Drosophila as compared to the injection of 1 mM NaCl, increased the percent lethals produced by 2000 R X-irradiation. The percent lethals in the NaCl groups was about 5% in four experiments while it was 6% to 10% in the NaF groups.

. Vogel (1973) stated that 1 mM F in 5% glucose solution fed to larva was a weak mutagen but that 12 mM F acted as a powerful anti-mutagen in combination with a known strong mutagen. The evidence for the weak mutagenic effect is very slim. In one of three experiments, there were three lethals; one in the control group and one in each of two exposure groups. With fluoride exposure alone, the egg-laying capacity was clearly depressed in most experiments, and the hatchability was generally, but not always, lower. Hence, Vogel concluded that fluoride had a sterilizing effect. Treatment with the trialkylating agent Trenimon alone, showed 13.1% lethals, but in combination with 12 mM F, 3.4% lethals occurred without a consistent difference in the number of eggs laid or in their hatchability. A second experiment showed 8.6 and 1.3%, respectively. Since at least 350 chromosomes were tested by phenotypic markers for each group in each experiment, these differences are clearly statistically significant.

Büchi and Bürki (1975) confirmed the protective effect of fluoride (20 mM) on the mutagenicity of Trenimon (20 mM) and showed that the protection was lost when Trenimon was fed to the males and the fluoride to the females.

Herskowitz and Norton (1962) showed a marked increase in the incidence of melanotic tumors in fluoride-treated groups of two strains of Drosophila. With the controls 0.0% and 7.1% occurred while treatment of the larvae with 1 to 30 mM fluoride caused a smooth dose response up to nearly 100% occurrence. The group sizes were 1500 so the numbers are highly significant statistically. Testimony at the September 1977 Congressional Hearings indicated that the melanotic tumors are like granulomas rather than neoplastic tumors (Burton, 1977).

Obe and Slacik-Erben (1973) reported an antimutagenic effect of fluoride. They found a 25 to 50% decrease in the number of chromosome breaks of human cells in vitro with each of three separate strong mutagens in combination with 1 mM fluoride. Slacik-Erben and Obe's (1976) attempt to clarify the role of sodium fluoride and the antimutagenic effects with Trenimon proves control data which shows no effect from 1 mM fluoride alone. The human cells in this case were lymphocytes stimulated with phytohaemagglutinin in which nucleotide incorporation and mitotic index were followed for over two days. The data curves for two experiments were averaged and plotted, hence, only a visual judgement of no effect can be made, and no estimate of the power of the test is given.

Jagiello and Ja-Shein (1974) exposed mammalian eggs to NaF and concluded that some changes were taking place. The earliest effect noted was clumping of the chromosomes at meiosis in cow oocytes exposed to 4.5 ppm (240 µM), the lowest concentration used. There is no indication whether the reading of smears was done with or without knowledge of treatment.

Chang (1967) showed that concentrations of fluoride which inhibited growth of corn seedlings (0.5 mM and higher) affected the nucleotide ratios. However, similar effects are seen when growth is inhibited by other means.

The above studies, with the exception of those by Bale and Hart (1973) where the dosage levels are questionable and the cow oocytes study, were all done with fluoride concentrations of 0.5 mM or higher. With 0.5 mM fluoride the physiological (mammalian) concentration of calcium (1.5 mM) makes the solution saturated in respect to calcium fluoride. Thus the cause of the above effects, assuming that they are real, may be a lowering of the calcium or metal ion content in solution rather than a direct effect of the fluoride on the genetic material. This distinction would be important since precipitation is clearly a threshold phenomenon, i.e., below supersaturation no effect would be expected; whereas mutagens which act directly on the genetic material may have no

threshold.   While the above studies suggest that F and HF
in high doses may be weak mutagens in Drosophila, they are
far from having demonstrated that fluoride has a mutagenic
potential in humans.   Further work clarifying the above questions would be interesting, however, with the large human
populations available for studying these questions with epidemiological techniques, the results in animals should not be
given much weight.

Fluoride and Cancer

Cancer in Animals

Taylor (1954) found a 9-10% decrease in the mean life
span in cancer-prone mice with drinking water at 1 ppm fluoride as compared to distilled water.   He concluded that fluoride accelerated the rate of tumor growth.   Taylor and Taylor
(1964, 1965) tested this hypothesis more directly with injected tumor material in animals and in eggs.   The conclusions,
however, are weakened by the very unusual dose-response relationships in all of the studies and by the lack of independent confirming studies.   In the 1954 study, one series of
experiments was done with commercial chow which contained 20
to 38 ppm F while in another series a low-F diet was used.   A
difference in survival occurred only between the animals receiving 0 and 1 ppm fluoride and no difference was noted between 1 and 10 ppm, which was also tested in the second series.   To be consistent with the expected serum fluoride concentrations (see Taves and Guy, this monograph) the effect
should have been seen between the 1 and 10 ppm F groups
rather than the 0 and 1 ppm groups.   The lack of an expected
dose response is more marked in the tumor growth experiments
where a much wider range in doses was used (0.05 to 25 ppm),
yet again the only difference noted was between the distilled
water and the 0.05 ppm groups.   Such findings generally indicate inadequate controls or the presence of bias.   The use of
distilled water for control rather than an equivalent sodium
chloride solution might be a cause of inadequacy in the controls.   The data of Flemming (1953) suggested a beneficial
effect from fluoride at 20 ppm in the drinking water of mice
with implanted tumors.   Bittner and Armstrong (1952, 1954)
found no carcinogenic effect with 5-10 ppm F in the drinking
water in the ZPC mice strain but did not measure survival
times.

Cancer in Humans

Early in 1975 Yiamouyiannis claimed that there is a
linkage between fluoridation of water and increased cancer
mortality rates in humans.   This claim was based on the sum
of the rates for nine specific cancer sites (7 for white

males and 2 for white females) for the ten largest U.S. cities with water supplies fluoridated for more than 12 years prior to 1970 and for the ten largest non-fluoridated U.S. cities. The source of the data was the age-specific cancer mortality rates for a 20-year period by site and county compiled by the National Cancer Institute (NCI) and published by the Department of Health, Education and Welfare (HEW) in 1974. There is clearly a difference between the two groups of cities, with the fluoridated ones having about 25/100,000 more cancer deaths than the non-fluoridated ones, but the data presented by Yiamouyiannis do not indicate whether the difference started with fluoridation.

In September 1975, Yiamouyiannis and Burk submitted data to NCI suggesting that there had been a change in cancer rates with time for fluoridated cities as compared to non-fluoridated ones. As confirmed by NCI, the average crude mortality rates diverge markedly after 1952 when the one group of cities initiated fluoridation (1952-1957). NCI noted, however, that Yiamouyiannis and Burk had failed to take into account different changes in demographic factors and age distributions that affect cancer rates. NCI used 1950 rates for the U.S. population as a whole to adjust the crude mortality rates for sex, race and age. The results are expressed as the ratio of observed deaths to expected deaths (standard mortality ratio; SMR). The SMR's are greater than unity for both sets of cities and greater in the fluoridated than in the non-fluoridated cities, but there is no change in these ratios with time from 1950 to 1970 (Fredrickson, 1976; Taves, 1977; Doll and Kinlen, 1977; Oldham and Newell, 1977).

Taves (1977) studies with the SMR's are the most extensive in that they include for each time period reported two years' deaths (each census year plus the year preceeding) and include a larger number of cities than the 20 which Yiamouyiannis and Burk studied. Taves' results are shown in Figure 1. In no case is there a statistically significant difference in time trends. Thus, the assertion that fluoridation is linked with an increase in overall cancer mortality rates does not hold.

The rates in fluoridated cities are higher only for a particular set of cities and the higher rates in these cities were present before fluoridation. The data for all 20 fluoridated cities and all 15 non-fluoridated cities are shown in Table 1.

For negative results like those described above it is important to assess the magnitude of the effect that would escape detection, i.e., confidence limits. For the Taves'

Table 1. SMRs of fluoridated and non-fluoridated U.S. cities
         including the two groups studied by Yiamouyiannis
         and Burk.

Source of SR*	No. of cities & F Status	SMR** 1950	SMR** 1960	SMR** 1970	Change 1950-1970	Diff.
U.S. 1950	15 no	1.1272	1.0899	1.1390	0.0118	+.0037
	20 yes	1.1962	1.1959	1.2117	0.0155	
U.S. 1950	10 no	1.1498	1.1303	1.1659	0.0160	+.0066
	10 yes	1.2302	1.2348	1.2529	0.0227	
15 NF+F*** (combined)	15 no	0.9344	0.9037	0.9469	0.0124	+.0034
	20 yes	0.9861	0.9848	1.0020	0.0158	
15 NF+F*** (combined)	10 no	0.9539	0.9378	0.9702	0.0163	+.0019
	10 yes	1.0120	1.0130	1.0301	0.0181	

  *Standard rates used for SMR calculations
 **Standard Mortality Ratios
***Non-fluoride + Fluoride

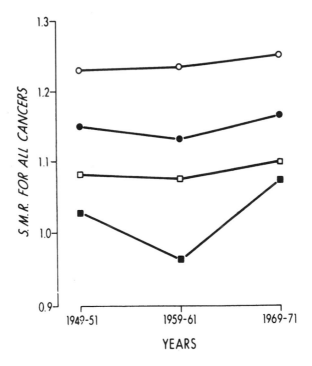

Figure 1. Standard mortality ratios for all cancers using
the average of two years' observed mortality. The ten
fluoridated (O) and ten non-fluoridated (●) U.S. cities
studied by Yiamouyiannis and Burk (1977). (■) represents
the five cities which they omitted from the 15 largest non-
fluoridated cities between 1950 and 1969. (□) represents
the next ten largest long-term fluoridated cities (Taves,
1977).

study, a 1.5% increase in cancer death rates would have been detected with 95% confidence based on the standard errors of the means.

In earlier but supplementary studies, Hoover et al. (1976) investigated the absolute differences in cancer rates in fluoridated and non-fluoridated areas. When demographic variables were taken into account, the differences became insignificant except for stomach cancer. Further regression analysis, controlling for specific high-risk ethnic groups, yielded a nonsignificant F value of 0.02 for females. The F* difference for males, however, still remained greater than expected by chance at the 95% confidence level.

A linkage between stomach cancer with fluoride would not be unreasonable because fluoride exists primarily as hydrofluoric acid, a highly penetrating and irritating chemical, in the acidic stomach. As noted above, hydrofluoric acid may be a weak mutagen in plants (Mohamed, 1969) and drosophila (Mohamed and Kemmer, 1970).

Okamura and Matsuhisa (1963) showed a correlation between stomach cancer and the fluoride content of rice and "miso". The fluoride values reported for food by Okamura are many times those expected in this country and are based on analytic methods which would not distinguish between organic and inorganic fluoride. So, even if there is a statistically significant correlation, it is not clear that fluoride ion is involved. Hirayama (1977) reported that stomach cancer rates in Japan were positively correlated with the amounts of hot tea and fish consumed and negatively correlated with the amount of milk drunk. Tea and fish have been reported to have higher levels of fluoride than other foods and milk would be expected to act as a binding agent and buffer to reduce the effective concentration of HF in the stomach.

Epidemiological studies in England fail to support the hypothesis that stomach or any other cancer is associated with fluoride intake (Kinlen, 1975; Royal College of Physicians, 1976).

Two other studies at NCI (Hoover et al., 1976) give additional information on the question of linkage of fluoride to stomach cancer. In the first, cancer data for all the U.S. counties in which at least 2/3 of the water supply was first fluoridated between 1950 and 1965 were grouped into 5-year intervals in order to study changes with time. Mortality data were compiled for 34 different cancer sites. None of the specific sites gave any indication of an increase in cancer following fluoridation; rather a possible decrease

---

*F here means Fisher test.

is suggested. The "other" category is of interest because
it is the only grouping which suggested a possible increase
for both sexes.

The second study compared naturally fluoridated and non-
fluoridated counties in Texas on the same basis as in the
previous one. The SMR's were more variable but there were
no consistent trends with increasing fluoride content except
for a possible decrease in the "other" category.

At the symposium (Feb., 1977) which is the basis for
this monograph, Yiamouyiannis presented some tabulations
which he claimed showed that the SMR calculations were unreliable and that there had been an increase in the cancer
rates for those over 45 years of age. These data have subsequently been published by Yiamouyiannis and Burk (1977) in
more detail. As further evidence for unreliability of the
SMR method, Yiamouyiannis and Burk quoted a "personal communication" for the increase in the U.S. cancer mortality rate
from 1950 to 1969 as being 13%. Since they found a 14% increase and the SMR studies only show a 2-3% increase in the
rates for the large cities in question, Yiamouyiannis concluded that there was an inherent error in the indirect (SMR)
method. However, the published increase in the age, sex and
race-adjusted cancer death rates for the U.S. was actually
4.2%, not 13% (from linear regression of data in Table C,
DHEW, 1974). Thus, it is their method which appears to be
unreliable.

Yiamouyiannis has a valid point when he says that more
attention should be given to the set of rates used with the
indirect method for calculating the expected number of
deaths. Therefore, the expected number of deaths were re-
calculated using rates derived from 15 of the 20 original
cities or corresponding counties which had complete break-
downs of cancer deaths by age, race and sex. (There were 3
cities for which rates for the corresponding counties had
to be used because of lack of available city data.) The
results of these calculations are shown in Table 1. They
show a difference in change of the cancer death rates of
only 0.19%. This difference is not statistically significant as the standard error of the difference in increase is
1.12%. However, it means that a 2% greater real increase in
the fluoridated cities compared to the controls can probably
be ruled out at the 95% confidence level and that an increase of 3% can certainly be ruled out.

Doll (1977) criticized Yiamouyiannis and Burk's claim
that there has been a significant increase in the cancer
mortality of the groups over 45 years of age. His criticism is based on their use of at least 20-year age groups
without sex and race adjustment. These adjustments cannot
be done directly at the present time because the data are not

available in sufficient detail. However, some aspects of
the controversy about the cancer death rates in fluoridated
and non-fluoridated U.S. cities can be checked indirectly.
Yiamouyiannis and Burk gave the ratio of the number of
people in the older half of the 20-year age groups to the
total number of people in order to determine any difference
in the ways in which the populations in the two groups of
cities aged. Their ratios and Taves' ratios derived from
calculations of comparable figures from the Census Bureau
data are shown in Table 2. I was unable to confirm 1/3 of
their calculations. The discrepancies are important in the
case of the 45-64 year groups for 1970 because different
ratios would give a spurious increase in cancer mortality
with fluoridation. The divergence in the composition is
3.5% or an apparent increase of 12/100,000 in the cancer
death rate in the fluoride cities. Therefore, their claim
of a 15/100,000 increase in that age group is not surprising
and is about what would be expected from the difference in
change in the age composition.

A more accurate and complete evaluation of the effect
of using the broad subcategories in the rate calculations
can be made indirectly. This evaluation is done by observ-
ing the effect of collapsing the data from 5-year age, sex
and race categories into the 20-year age groups and disre-
garding sex and race. The populations of the five-year age,
sex and race subcategories for the fluoridated and non-
fluoridated cities are multiplied by a single set of cancer
rates to obtain an expected number of deaths. These expected numbers of deaths and the respective populations are
summed in the broad categories and the respective rates calculated by dividing the similarly collected population
figures into the expected deaths. Table 3 shows the results
of these calculations; 7.4/100,000 of Yiamouyiannis and
Burk's increase in the 45-64 year age group but none of the
increase in the 65 year + age group can be explained on the
basis of the changes in age, race and sex composition.
Assuming that the changes found by Yiamouyiannis and Burk in
cancer rates for the less than 45 year age group were identical in the fluoridated and non-fluoridated cities, the
over-all unexplained increase in cancer mortality would be
2.55%. This exceeds the increase likely to have been missed
by the indirect method, therefore, the discrepancy appears
to be real. Since the errors noted above were found only
with accessible data, it is possible that further errors
will be found if detailed data are made available. In any
case, the risk ratios are too small to substantiate a causal
relationship without an exhaustive study on alternative explanations and production of convincing animal data.

Table 2.   Age distribution within age groups for fluoridated
(F) and non-fluoridated (NF) cities.

(Yiamouyiannis' [Y] vs Taves' [T] Calculations)					
	1950		1960		1970
	Y	T	Y	T	Y	T
F (55-65/45-64)	.43	.435	.45	.450	.48	.471
NF (55-64/45-64)	.43	.434	.45	.446	.47	.456
F (74+/65+)	.29	.286	.31	.305	.37	.368
NF (75+/65+)	.29	.306	.34	.337	.38	.387

Table 3.   Rates per 100,000 in broad subcategories; detailed
(age span [5 yr], race, and sex) rates are identical for populations of 10 fluoridated (F) and
10 non-fluoridated (NF) cities.

1950	0-25	25-44	45-64	65+
F	7.41	43.6	340.6	991.8
NF	7.47	43.4	337.1	992.6
1960				
F	6.96	47.6	348.8	983.4
NF	7.10	46.2	343.1	989.4
1970				
F	6.88	47.0	360.3	993.4
NF	7.15	44.4	349.4	993.9
Difference 1950-1970				
F	-0.53	+3.3	+19.7	+1.6
NF	-0.32	+1.0	+12.3	+1.3
Excess	-0.21	+2.3	+ 7.4	+ .3

It will be relatively easy to find other differences between the fluoridated and non-fluoridated cities which might provide alternative explanations. For instance, the total mortality rates (SMRs) for these two groups of cities have decreased 17% and 19% in the period from 1950 to 1970 for the non-fluoridated and fluoridated cities, respectively. The decreases in SMRs for heart disease are 17% and 24%, respectively (Taves, 1978). In the face of these larger and opposite changes occurring in the population under consideration, it would have to be shown that the smaller differences being claimed by Yiamouyiannis and Burk are not an artifact or an indirect consequence of the larger changes.

Concluding Remarks

The data used to support the claims that fluoridation causes adverse effects in humans are not convincing. The only "statistically" significant difference which is as yet unexplained in human cancer data is for one group (over 65 years of age) which has been selected from a group of four comparisons. The p value for this age group was borderline when taken as a single independent observation. The appropriate p value needs to be calculated on the basis of a set of 4 observations (Snedecor and Cochran, 1967) but will almost certainly be greater than 0.05. The studies using the SMR for total cancer mortality rule out an over-all increase of more than 1% to 2%. Further resolution or reduction in the margin of error can be achieved when data for further census years are available for analysis. However, it should be remembered that while the margin of possible error can be reduced, theoretically, absolute proof of safety cannot be attained. Also, the remaining possible risks and the benefit are in different units so that a comparison will remain subjective, and science, per se, cannot make the decision.

Acknowledgments

Portions of this paper were adapted from <u>Drinking Water and Health</u>, pages 377-379, 385, 388, and 390-94, with the permission of the National Academy of Sciences, Washington, D.C. Editorial assistance from NAS is appreciated, as is constructive criticism given by Dr. Albert W. Burgstahler, University of Kansas and Dr. Frank A. Smith, University of Rochester; however, their assistance should not be construed as agreement on all points.

Author's note: As this book goes to press, it has been pointed out that the discrepancy noted in Table 2 is probably explainable by the difference between using weighted versus unweighted values in these calculations.

References

Abid, A.A. (1967) The effects of sodium fluoride. Texas A & M University Thesis for Ph.D.

Armstrong, W.D. (1954) The scientific literature relating to fluoride and cancer. U.S. House of Rep. Comm. on Interstate and Foreign Commerce, pp. 306-310.

Bale, S.S. and Hart, G.E. (1973a) Studies on the cytogenetic and genetic effects of fluoride on barley. I. A comparative study of the effects of sodium fluoride and hydrofluoric acid on seedling root tips. Can. J. Genet. Cytol., 15;695-702.

Bale, S.S. and Hart, G.E. (1973b) Studies on the cytogenetic and genetic effects of fluoride on barley. II. The effects of treatments of seedling coleoptiles with sodium fluoride. Can. J. Genet. Cytol., 15; 703-712.

Berry, W.T.C. (1958) A study on the incidence of mongolism in relation to the F content of H_2O. Am. J. Ment. Defic., 62; 634-636.

Bittner, J.J. and Armstrong, W.D. (1952) Lack of effects of fluoride ingestion on longevity of mice. J. Dent. Res., 31;495.

Büchi, R. and Bürki, K. (1975) The origin of chromosome aberrations in mature sperm of Drosophilia: Influence of sodium fluoride on treatments with Trenimon and 1-phenyl-3, 3-dimethyltriazene. Archiv. Genetik., 48;59-67.

Burgstahler, A.W. (1975) Fluoride and Down's syndrome (Mongolism). Fluoride, 8;1-11.

Burgstahler, A.W. (1977) Subcomm. on Intergovmntl. Relations and Human Resources. U.S. House of Rep. Comm. on Govemntl. Operations; p. 305.

Burton, G.J. (1977) Melanotic and other tumors in relation to sodium fluoride and other substances in the fruit fly Drosophila melanogaster. U.S. House of Rep. Comm. on Govmntl. Operations; pp. 488-489.

Cavagna, G. and Bobbio,G. (1970) Contributo allo studio delle caratteristiche chimico-fisiche e degli effetti biologici degli effluenti di una fabbrica di alluminio. Med. Lavoro, 61;69-101.

Chang, C.W. (1968) Effect of fluoride on nucleotides and ribonucleic acid in germinating corn seedling roots. Plant Physiol., 43;669-674.

Colombini, M. et al. (1969) Observations on fluorine pollution due to emissions from an aluminum plant in Trentino.

Fluoride, 2;40-48.

Cook, H.A. (1970) Fluoride intake through tea in British
children. Fluoride, 3;12-18.

Cook, H.A. (1972) Crippling arthritis related to fluoride
intake (case report). Fluoride, 5;209-213.

DHEW (1974) Mortality trends for leading causes of death;
United States, 1950-69. Vital and Health Statistics,
Series 20(16), Natl. Ctr. Hlth. Stat. Rockville, Md.,p.11.

Doll, R. (1977) Fluoridation and Cancer. Lancet, 2;296.

Doll, R. and Kinlen, L. (1977) Fluoridation of water and
cancer mortality in the USA. Lancet, I;8025, 1300-1302.

Douglas, T.E. (1957) Fluoride dentifrice and stomatitis.
Northwest Med., 56;1037.

Erickson, J.D. et al., (1976) Water fluoridation and congenital malformations: no association. J. Am. Dent. Assoc.,
93;981-984.

Ericsson, Y. (1970) Introduction in: Fluorides and Human
Health, WHO Monograph No. 59, Geneva, pp. 13-16.

Feltman, R. (1956) Prenatal and postnatal ingestion of
fluorides: a progress report. Dent. Digest, 62;353-357.

Feltman, R. and Kosel, G. (1961) Prenatal and postnatal
ingestion of fluorides - 14 years. Final report. J. Dent.
Med., 16;190-198.

Finn, R. and Cohen, H.N. (1978) "Food Allergy": Fact or
Fiction? Lancet, 1;426-428.

Fleming, H.S. (1953) Effect of fluorides on the tumors S37
after transplantation to selected locations in mice and
guinea pigs. J. Dent. Res., 32;646.

Frederickson, D.S. (1976) Letter (February 6) to Hon. L.L.
Delaney, U.S. House of Representatives.

Fry, B.W. et al. (1973) Fluorometabolite of methoxyflurane.
Anesthesiology, 38;38.

Goodman, L.S. and Gilman, A. (1975) The Pharmacological
Basis of Therapeutics, 5th ed., MacMillan, New York.

Green, D.M. (1964) Pre-existing conditions. Placebo reactions and "side effects". Ann. Intern. Med., 60;255-265.

Grimbergen, G.W. (1974) A double-blind test for determination of intolerance to fluoridated waters. Fluoride, 7;
146-152.

Handbook of Chemistry (1976-77) Ed., R. O. Weast, CRC Press,
Cleveland, p. B-160.

Herskowitz, I.H. and Norton, I.L. (1963) Increased incidence of melanotic tumors in two strains of Drosophila melanogaster following treatment with sodium fluoride. Genetics, 48;307-310.

Hirayama, T. (1977) Changing patterns of cancer in Japan with special reference to the decrease in stomach cancer mortality. In: Origins of Human Cancer, eds., Hiatt, H.H., Watson, J.D. and Winsten, J.A., Cold Spring Harbor, pp. 55-75.

Hodge, H.C. and Smith, F.A. (1965) In: Fluorine Chemistry, ed. Simmons, J.H., Academic Press, New York, Vol. IV, p. 135.

Hodge, H.C. and Smith, F.A. (1977) Occupational fluoride exposure. J. Occup. Med., 19;12-39.

Hoover, R.N. et al. (1976) Fluoridated drinking water and the occurrence of cancer. J. Natl. Cancer Inst., 57; 757-768.

Husdan, H. et al. (1976) Serum ionic fluoride: normal range and relationship to age and sex. Clin. Chem., 22;1884-1888.

Jagiello, G. and Ja-Shein, L. (1974) Sodium fluoride as potential mutagen in mammalian eggs. Arch. Environ. Hlth., 29;230-235.

Kinlen, L. (1975) Cancer incidence in relation to fluoride level in water supplies. Brit. Dent. J., 138;221-224.

Mohamed, A.H. et al. (1966a) Cytological reactions induced by sodium fluoride in Allium cepa root tip chromosomes. Can. J. Genet. Cytol., 8;241-244.

Mohamed, A.H. et al. (1966b) Cytological effects of hydrogen fluoride on tomato chromosomes. Can. J. Genet. Cytol., 8;575-583.

Mohamed, A.H. (1968) Cytogenic effects of hydrogen fluoride treatment in tomato plants. J. Air Pollution Control Assoc., 18;395-398.

Mohamed, A.H. (1969) Cytogenic effects of hydrogen fluoride on plants. Fluoride, 2;76-84.

Mohamed, A.H. (1970) Chromosomal changes in maize induced by hydrogen fluoride. Can. J. Genet. Cytol., 12;614-620.

Mohamed, A.H. (1971) Induced recessive lethals in second chromosomes of Drosophila melanogaster by hydrogen fluoride. In: 2nd Internat. Clean Air Cong., eds., Englung, H.M. and Berry, W.T., Academic Press, New York, pp. 158-161.

Mohamed, A.H. and Kemner, P.A. (1970) Genetic effects of hydrogen fluoride on Drosophila melanogaster. Fluoride, 3;192-200.

Mohamed, A.H. and Chandler, M.E. (1977) Cytological effects of sodium fluoride on mice. U.S. House of Rep. Comm. on Govmntl. Operations; Subcomm. on Intergovmntl. Relations and Human Resources, pp. 42-48.

Mohamed, A.H. (1977) Cytogenic effects of hydrogen fluoride gas on maize. Fluoride, 10;157-165.

Mouftah, S.P. (1968) Mitotic aberrations in Vicia fabia L chromosomes induced by sodium fluoride. Texas A & M Univ. Thesis for Ph.D.

Mukherjee, R.N. and Sobels, F.H. (1968) The effects of sodium fluoride and iodoacetamide on mutation induction by x-irradiation in mature spermatozoa of Drosophila. Mutat. Res., 6;217-225.

NAS-NRC Committee on Anesthesia (1971) Statement regarding the role of methoxyflurane in the production of renal dysfunction. Anesthiol., 34;505-509.

Needleman, H.L., Pueschel, S.M. and Rothman, K.J. (1974) Fluoridation and the occurrence of Down's syndrome. N.Engl. J. Med., 291;821-823.

Obe, G. and Slacik-Erben, R. (1973) Suppressive activity by fluoride on the induction of chromosome aberrations in human cells and alkylating agents in vitro. Mutat. Res., 19;369-371.

Okamura, T. and Matsuhisa, T. (1963) The fluorine content in favorite foods of Japanese. Nippon Sakamotsu Gakkai Kiji. 32;132-138.

Oldham, P.D. and Newell, D.J. (1977) Fluoridation of water supplies and cancer - a possible association? Appl. Statist., 26;125-135.

Olivo, R. et al. (1968) Annali Sanita Pubblica, 19;1659.

O'Meara, W.F. (1968) Fluoride administration in single daily dose: survey of its value in prevention of dental caries. Clin. Pediat., 7;177-184.

Petraborg, H.T. (1974) Chronic fluoride intoxication from drinking water. Fluoride, 7;47-52.

Quissell, D.O. and Suttie, J.W. (1972) Development of a fluoride-resistant strain of L cells: membrane and metabolic characteristics. Am. J. Physiol., 223;596-603.

Rapaport, I. (1956) Contribution a l'etude du mongolisme. Role pathogenique du fluor. Bull. Acad. Natl. Med. (Paris) 140;529-531.

Rapaport, I. (1959) Nouvelles recherches sur le mongolisme. A propos du role pathogenique du fluor. Bull. Acad. Natl. Med. (Paris), 143;367-370.

Rapaport, I. (1963) Oligophrenie mongolienne et caries dentaires. Rev. Stomatol. (Paris), 46;207-218.

Repaske, M. and Suttie, F.W. (1978) Effects of fluoride on cultured cells: Metabolism and fluoride resistance. This monograph.

Rich, C. et al. (1964) The effects of sodium fluoride on calcium metabolism of subjects with metabolic bone diseases. J. Clin. Invest., 43;545-556.

Royal College of Physicians (1976) Fluoride, Teeth and Health. Pitman Medical, London.

Schlesinger, E.R. (1970) Health studies in areas of the USA with controlled water fluoridation. In: Fluorides and Human Health, WHO Monograph No. 59, Geneva, pp. 305-310.

Shambaugh, Jr., G.E. and Siva Sundar, V.S. (1969) Experiments and experiences with sodium fluoride for inactivation of the otosclerotic lesion. Laryngoscope, 79;1754-1764.

Shea, J.J., Gillespie, S.M. and Waldbott, G.L. (1967) Allergy to Fluoride. Ann. Allergy, 25;388-391.

Slacik-Erben, R. and Obe, G. (1976) The effect of sodium fluoride on DNA synthesis, mitotic indices and chromosomal aberrations in human leukocytes treated with Trenimon in vitro. Mutat. Res., 37;253-266.

Snedecor, G.W. and Cochran, W.G. (1967) Statistical Methods. Iowa State Univ. Press, Ames, Iowa, 6th ed., pp. 271-276.

Steinegger, S. (1969) Endemic skin lesions near an aluminum factory. Fluoride, 2;37-39.

Tao, S. and Suttie, J.W. (1976) Evidence for a lack of an effect of dietary fluoride level on reproduction in mice. J. Nutr., 106;1115-1122.

Taves, D.R. (1977) Fluoridation and cancer mortality. In: Origins of Human Cancer, Cold Spring Harbor Conf. on Cell Proliferation, eds., H.H. Hiatt, J.D. Watson and J.A. Winsten, 4;357-366.

Taves, D.R. (1978) Fluoridation and mortality due to heart disease. Nature, 272;361-362.

Taves, D.R. and Guy, W.S. (1978) Distribution of fluoride among body compartments. This monograph.

Taylor, A. (1954) Sodium fluoride in the drinking water of mice. Dent. Digest, 60;170-172.

Taylor, J.M. et al. (1961) Toxic effects of fluoride on the rat kidney. II. Chronic effects. Toxicol. Appl. Pharmacol., 3;290-314.

Taylor, A. and Taylor, N.C. (1964) Effect of sodium fluoride on tumor growth. Can. Res., 24;751.

Taylor, A. and Taylor, N.C. (1965) Effect of sodium fluoride on tumor growth. Proc. Soc. Exp. Biol. Med., 119;252-255.

Temple, P.J. and Weinstein, L.H. (1976) Personal communication, November 23.

U.S. Bureau of the Census. 1950, 1960, 1970. Census of the population by states. U.S. Government Printing Office, Washington, D.C.

VanDyke, R.A. (1978) Fluoride from anesthetics and its consequences. This monograph.

Vogel, E. (1973) Strong antimutagenic effects of fluoride on mutation induction by Trenimon and 1-phenyl-3, 3-dimethyl-triazene in Drosophila melanogaster. Mutat. Res., 20; 339-352.

Waldbott, G.L. (1962) Fluoride in clinical medicine. Int. Arch. Allergy Appl. Immunol. 20(suppl. 1);29.

Waldbott, G.L. and Steinegger, S. (1973) New observations on "Chizzola" maculae. Proc. Third Internatl. Clean Air Cong., Dusseldorf.

Yiamouyiannis, J.A. (1975) A definite link between fluoridation and cancer death rate. Natl. Hlth. Fed., typed manuscript, March 25.

Yiamouyiannis, J.A. (1977) Fluoridation and Cancer. U.S. House of Rep. Comm. on Govmntl. Operations; Subcomm. on Intergovmntl. Relations and Human Resources, pp. 61-72.

Yiamouyiannis, J.A. and Burk, D. (1977) Fluoridation and cancer, age-dependence of cancer mortality related to artificial fluoridation. Fluoride, 10;102-125.

Zanfagna, P.E. (1976) Allergy to Fluoride. Fluoride, 9;36-41.

Printed and bound by CPI Group (UK) Ltd, Croydon, CR0 4YY

23/10/2024

01778241-0017